P9-EKY-378

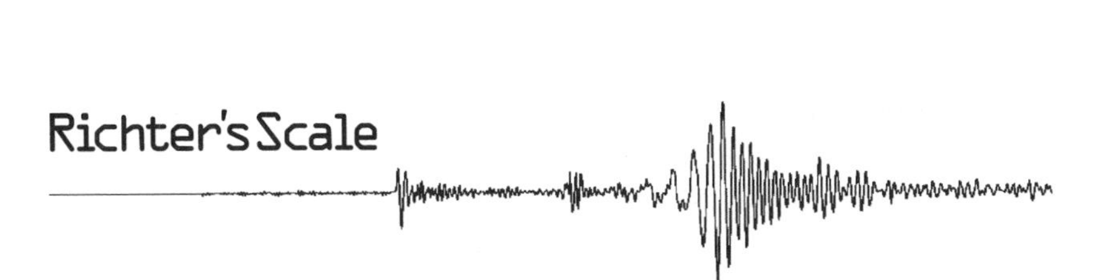
Richter's Scale

Richter's Scale

MEASURE OF AN EARTHQUAKE • MEASURE OF A MAN

SUSAN ELIZABETH HOUGH

PRINCETON UNIVERSITY PRESS · PRINCETON AND OXFORD

Published by Princeton University Press, 41 William Street, Princeton, New Jersey 08540
In the United Kingdom: Princeton University Press,
3 Market Place, Woodstock, Oxfordshire OX20 1SY

Library of Congress Cataloging-in-Publication Data

Hough, Susan Elizabeth, 1961–
Richter's scale : measure of an earthquake, measure of a man / Susan Elizabeth Hough.
p. cm.
Includes bibliographical references and index.
ISBN-13: 978-0-691-12807-8 (cloth : alk. paper)
ISBN-10: 0-691-12807-3 (cloth : alk. paper)
1. Richter, C. F. (Charles Francis), 1900– 2. Seismologists—United States—Biography.
3. Richter scale. 4. Earthquakes. I. Title.
QE22.R475H68 2007
551.22092—dc22
[B] 2006016480

British Library Cataloging-in-Publication Data is available
This book has been composed in Minion typeface.
Printed on acid-free paper. ∞
pup.princeton.edu
Printed in the United States of America

10 9 8 7 6 5 4 3 2 1

For my mother,

BARBARA SMITH,

who puts up with me;

and my father,

JERRY HOUGH,

who taught me the importance of
original sources (in 10th grade);

and my stepmother,

JEAN CRAWFORD,

who puts up with my father.

CONTENTS

PREFACE

> If I have anything to offer, it must be something individual and peculiar to myself.
>
> —*Charles Frances Richter, May 26, 1937*

> Charles Richter contributed a lot more than the Richter Scale to humanity—he contributed himself,
>
> —*Wanda Tucker,* Pasadena Star News, *October 3, 1985*

In spring of 1981, I turned twenty; Charles Richter had just the previous year reached his milestone eightieth birthday. An undergraduate student at UC Berkeley, I declared a major in geophysics—with a specific interest in seismology—just as Charles Richter slipped away from the field, having been one of its brightest luminaries for nearly a half century. Richter spent the last few years of his life away from not only the public eye but largely also away from the eyes of those who had known him. I never had a chance to meet him; he died on September 30, 1985. His wife Lillian had died in 1972. The couple had no children; Richter's family in particular was limited to a sprinkling of distant cousins whom he never met, and probably never knew existed. Moreover, throughout his long professional career, Richter worked closely with very few students, leaving him with few professional progeny and no descendents in the biological sense. Even Charles's and Lillian's longtime home in Pasadena had vanished, claimed by eminent domain in the late 1960s to make way for the Foothill Freeway.

My career in seismology brought me from the northern end of California to the southern end, where I completed a Ph.D. in Earth Sciences at the Scripps Institution of Oceanography in 1987. From there my family and I moved to New York, where my husband, a biochemist, and I spent our post-doc years. In 1992 a job with the U.S. Geological Survey in Pasadena brought me back to California. Here my professional orbit overlapped with Richter's, at least in space: the USGS Pasadena office is on the Caltech campus, directly across the street from the building that has housed the Seismological Laboratory since 1974.

During my years with the USGS I became part of the legacy of observational seismology that traces its roots directly to Charles Richter and his colleagues—pioneers of the modern field of inquiry. I sometimes deploy portable seismometers to record aftershocks and other earthquakes. These instruments bear little resemblance to, yet trace their heritage directly back to, the first portable seismometers that Richter's colleagues designed and that he himself deployed after the large Southern California earthquakes of his day. I analyze seismograms recorded on both portable seismometers and the permanent monitoring network, the latter again a direct descendent of developments made by Charles Richter and his colleagues. Some of my colleagues at the Seismo Lab, as it is informally known, and elsewhere were Richter's colleagues twenty-five or more years ago.

As it turns out, Charles Richter and I have more in common than I guessed before embarking on this project. In particular, we share (or shared; one struggles here with tense) the same favorite local wilderness destination (Sequoia National Park) as well as a lifelong interest in writing: not only technical material but also science articles for nonspecialist audiences. We have both, I learned, published articles in the same popular magazine, *Natural History.* Richter's outlet of choice for verbal expression was different from mine, however: he wrote poetry, not as an occasional diversion but as a passion to rival his intense interest in earthquakes.

My previous writing projects have been more standard sorts of efforts: a book about earthquake science for a nonspecialist audience, a guidebook to California's faults (of the geological sort), a consideration of past and future earthquakes on an urban planet. I certainly never imagined myself as a biographer, although biographies such as Sylvia Nasar's *A Beautiful Mind* and Simon Winchester's *The Professor and the Madman* have been some of my favorite books in recent years.

In the midst of a conversation about future books with a writer friend, I made an offhand remark that it seemed strange nobody had written a biography of Charles Richter. Richter is, I commented, not only the one seismologist everybody knows, but also an individual known to have been decidedly iconoclastic.

I further observed that, while it seemed that someone ought to write the story, it couldn't be me. Aside from the fact that I am not a biographer, a "tell-all" book about a seismologist of iconic stature is not the sort of book a person should write if she intends to remain a member in good standing of the seismological community. Another concern seemed secondary at the time: that Richter had been so private that it might not be possible to put together any semblance of a life story. And there the matter lay, until, gradually, the idea began to resurface, and to grow on me. When one walks into the Caltech bookstore one finds an entire Einstein section, a row of books on Feynman, more than a few on Oppenheimer—and zero on the scientist whose name recognition might match that of any other Caltech scientist past or present. (Einstein in fact spent just three semesters at Caltech.)

Then one day I ventured down into the bowels of Caltech's Beckman building, around several corners and past many doors with signs warning, "Danger! Laser Radiation!" and arrived at the Caltech archives, to whom Richter left his papers. *All* of his papers. From the first day that I began to read the story, I knew that I wanted to tell it. I learned that a proper biography of Charles Francis Richter would not be a gossipy, "tell-all" story at all, but rather an enormously compelling story of an enormously compelling man: brilliant, yet tormented by demons that very nearly got the best of him; iconoclastic even by scientists' standards.

The story does include its more personal elements. As has been known for years in the seismological community, Richter was an avid nudist. As has been less well known, his unconventional, sometimes turbulent marriage was punctuated by a number of outside entanglements with other women, some of them quite serious. Why write of such things, some people ask: they were clearly private matters. And yet they are part of who Charles Richter was, and therefore part of the story. To tell that story—to paint as full a portrait of the man as is possible—requires a consideration of the whole, not just the socially acceptable parts. Although it might have been a life lived outside of usual societal—not to mention conventional religious—dictates, it was a life of honor, of integrity, of deep compassion and consideration for his fellow man (and woman). To surviving members of Lillian's family—wonderful individuals, all—who might wonder why skeletons couldn't remain in closets, I offer this: if God made us all so wonderfully different, surely He has some appreciation for those among us who break the mold.

Gradually I came to feel a tremendous sense of personal responsibility with this story. It is remarkable that it had not been written before. Richter left his papers to the Caltech archives, not only professional material but highly personal material as well, including mountains of his verse and other writings. It was his decision that this material—including the bits that will raise many eyebrows—should be preserved; he made a contractual agreement with the archives when he was in his seventies, very much of sound mind. One can only conclude that he wanted the story to be told. For whatever reason, or maybe for no reason, the job of telling it has fallen to me. Researching and writing Richter's story has, in a sense, given me the privilege of being the first person to *read* Richter's story: to put the pieces together, perhaps (and yes, I do recognize the arrogance of the thought) even more completely than anyone who knew him in person. It has been the journey of a lifetime.

Richter's Scale

CHAPTER 1

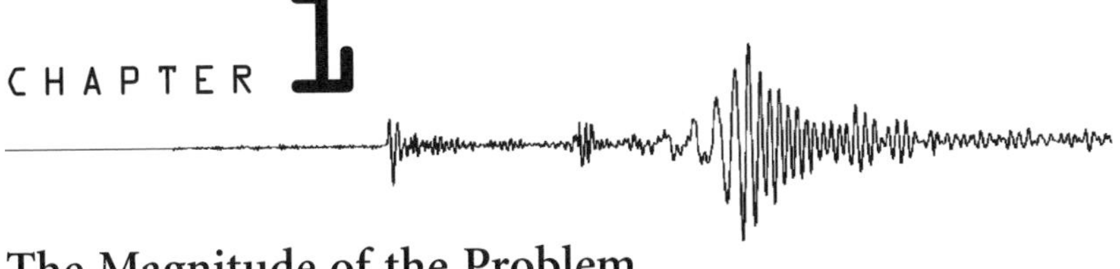

The Magnitude of the Problem

. . . it is good to have measured myself, to recognize my limitations.
—*Charles Richter, journal entry, June 20, 1926*

PIONEER settlers on the westernmost American frontier most likely settled into bed comfortably on the night of December 15, 1811. The Mississippi River valley had been enjoying an Indian summer: nighttime temperatures hovered near forty-five degrees Fahrenheit. The quiet and comfort of the settlers' slumber would, however, be shattered a few short hours later by the most portentous seismic disruption that had ever been witnessed by people who called themselves Americans. An eyewitness close to the seat of the disturbance described the scene around him: trees "bending as if they were coming to the ground—again, one rises as if it were to re-instate, and bending the other way, it breaks in twain, and comes to the ground with a tremendous crash." Astonishingly, this account described not the mainshock that shattered the still night around two-thirty, but rather its largest *aftershock*, which struck near dawn. And the account is remarkable. Rarely do trees snap in two even in strong earthquakes; it happens only in the most severe shaking the earth can dish up. (Even severe hurricane winds will generally yank a tree out of the ground rather than snap it in two.)

Between the wee hours of the morning of December 16, 1811, and February 7 of the following year, the midcontinent would be rocked by four enormously powerful earthquakes—the initial mainshock and its largest aftershock as well as subsequent large shocks on January 23 and February 7, 1812, and many thousands of smaller aftershocks. Waves from the largest shocks rippled outward with gusto through the midcontinental region. Although newspaper accounts reveal that these waves did not, as some still like to report, ring church bells in Boston, they did plenty else. Soft sediments along the Mississippi River gave way; the waters of the mighty river sloshed like waves in a bathtub, even reversing course for a brief time following the February 7 quake. Farther afield the attenuated waves still had enough power to crack brick walls in St. Louis, topple chimneys in Louisville, swing cabinet doors in Cincinnati, and damage plaster walls as far away as coastal South Carolina. The bell of St. Philip's church in Charleston, South Carolina, was set into motion. The enormous disruption and reach of these earthquakes led many to believe—even as late as the end of the twentieth century, after other large quakes had struck California—that the largest New Madrid earthquakes, as they came to be known, were the largest temblors to ever visit the contiguous United States.

The 1906 San Francisco earthquake caused substantially more loss of life and property damage, yet its overall effects seem to pale in comparison with those of the New Madrid earthquakes. According to pioneering geologist Grove Karl Gilbert, who investigated the effects of the 1906 earthquake, at distances of just twenty miles from the surface break, only an occasional chimney was overturned; by seventy-five miles the waves had lost their destructive punch altogether. (Because the rupture was several hundred miles long, damage extended over this full distance lengthwise along the San Andreas Fault.) Compare that with the collapsing riverbanks, reversing rivers, and damage as far as six hundred miles away caused by the New Madrid earthquakes.

But how do you measure an earthquake? This seemingly simple question proves complicated beyond all expectation. Prior to the 1930s, the best scientific minds in the world had no answer. In fact, they had barely begun to pose the question. Earlier scientists had devised methods to rank the severity of shaking based on its effects at different locations, but never a way to size up the temblors themselves. The difference is fundamental, essentially the same as that between the apparent brightness of a star in the nighttime sky here on Earth and its inherent luminosity—how brightly it shines up close. The effects of an earthquake depend on not only the distance from the fault to any given site, but myriad other factors as well. Seismic waves travel much more efficiently through the older and less complex rocks that make up the crust of central North America than they travel in California. Thus an earthquake of a given magnitude will pack a much greater punch in the former region than in the latter. And relative to the places most Californians now live, early American settlers were clustered in proximity to waterways, where earthquake shaking is significantly amplified by loose and wet sediments. The New Madrid earthquakes therefore hit eyewitnesses especially hard. As modern seismologists first endeavored to estimate the relative size of these earthquakes and of the 1906 earthquake, their results seemed reasonable: the San Francisco quake was the smaller.

I will leave a longer discussion of the New Madrid sequence for a later chapter and focus on the more fundamental question, how do you measure an earthquake? Nowadays, of course, any basic seismology textbook explains how earthquakes are measured, although basic texts still offer far more simplifications than subtleties of the methodology. It might surprise many readers to learn that these subtleties still rear their pesky heads within the corridors of research science, and not only for historical earthquakes for which we have limited data. When a powerful earthquake struck near Sumatra on the day after Christmas of 2004, global earthquake monitoring networks reported an initial magnitude estimate of 8.1; the estimate rose to 8.5 within hours and again to a staggering 9.0 a few hours later. Magnitude 9 earthquakes are, mercifully for us all, rare events: on average they strike perhaps once every few decades. (Although, one must note, the earth is not bound by averages: the 1960 Chilean and 1964 Alaska earthquakes were magnitude 9.5 and 9.2, respectively, still the largest two earthquakes in modern times, and a scant four years apart.) The low initial magnitude estimates for Sumatra reflected the fact that, while sophisticated global earthquake monitoring networks have been developed in recent decades, these networks and systems had never before been put through their paces with an earthquake of such portentous magnitude. The low initial estimates may have contributed to an underestimation of the tsunami potential: the bigger the earthquake, the larger the volume of water it can displace. (In retrospect, however, even a magnitude of 8 should have been sufficient to sound the alarms, had alarm systems been in place.) As the world watched with horror, the earthquake unleashed a deluge of biblical proportions, giant waves that swept over the coasts of Indonesia and Thailand before traveling the full width of the Bay of Bengal to inundate the coasts of southern India, Sri Lanka, and the Maldives.

Weeks after the earthquake a team of respected seismologists, Seth Stein and Emile Okal, began to circulate the results of their detailed analysis, which yielded an even higher magnitude estimate: 9.3. Although the analysis of Stein and Okale appears to have been beyond reproach, and was soon corroborated by other researchers, many seismologists expressed a reluctance to adopt the value because it had been estimated with a kind of data that are not available for earlier great earthquakes, specifically those in 1960 and 1964. Were equivalent data available for the earlier quakes, their magnitude estimates would very likely increase as well. Since one cannot make these calculations, most scientists reasoned, better to provide consistently determined estimates for all three temblors and thus an accurate description of their relative sizes. The alternative would be to upgrade Sumatra to a 9.3 while, by necessity, leaving Alaska at a 9.2, even though most scientists strongly suspect (but cannot prove) that Alaska was the larger of the two.

One begins, perhaps, to get an inkling of the magnitude of the problem. The business of sizing up earthquakes has been a surprisingly complex journey of discovery within the seismological community—one that traces its earliest roots to the years before modern seismometers were invented but began to gain traction only with Charles Richter's pioneering efforts in the 1930s. Earthquakes are, as it turns out, not only unruly but also terribly complicated beasts, the nature of which scientists began to understand only in the closing years of the nineteenth century. This is perhaps a surprising part of the story: prior to 1900, give or take a few years, scientists did not understand that an earthquake is, fundamentally, the abrupt movement of large parcels of the earth's crust along mostly flat surfaces known as faults. Prior to the closing years of the nineteenth century, scientists had advanced any number of other theories to explain the fundamental nature of earthquakes, for example underground explosions or electrical disturbances.

Once one understands that earthquakes involve motion along faults, one understands why size matters. That is, although earthquakes are generally named after the city they most heavily impact, they in fact occur along extended patches of faults, and the bigger the patch, the bigger the earthquake. Thus did the catastrophic 2004 Sumatra quake involve a patch of fault whose width was approximately 150 kilometers and whose length reached a staggering 1,500 kilometers. A map of California provides a useful sense of scale: the state measures about 1,000 kilometers from north to south. This one earthquake, then, unzipped a segment of fault equivalent to the full length of California, stem to stern, and then some. That's one big earthquake.

One returns again, however, to the question: how big is big? The previous paragraphs provide the answer (9.0), and explain that this reflects the size of the fault, but what does "magnitude 9" mean? Some quantities in science are relatively simple in the scheme of things. Take temperature, for example. Temperature fundamentally indicates the average energy of molecules in a substance. Nobody but a scientist thinks of temperature this way, but temperature is a familiar metric, one that can be reported as a simple numerical reading from a simple mechanical scale. The scale is moreover set, or calibrated, in a way that is easy to explain, in particular the Celsius scale: at sea level on planet Earth, water freezes at 0 degrees C and boils at 100 degrees C. On the Celsius scale, 100 is thus a physically meaningful number.

So, too, are we able to measure any number of other things: mass, force, speed. Such estimates become complicated only when bodies travel at close to the speed of light, in which case Einstein's theory of relativity begins to do strange things to the universe we know and love. But short of this most extreme situation, simple mechanical devices and scales suffice to measure quantities like force and mass. Scientists speak of these quantities as *parameters*: well-defined quantities that can be measured. An earthquake, on the other hand, is not a fundamental parameter as much as a process. The difference between measuring the mass of an object and the magnitude of an earthquake is a little like the difference between measuring the speed of one car and measuring the traffic on the New Jersey turnpike. The speed of one car is a parameter; the traffic on the turnpike is . . . something else.

Later chapters will delve further into both the nature of earthquakes and the first scale developed by scientists to measure them: Richter's scale. This book is, however, not only a story about earthquakes or the Richter scale, but also the story of Charles Francis Richter, the man. Richter is, even today, the only seismologist living or dead whose name is a household word throughout the world—a measure (so to speak) of immortality that stems directly from the scale that bears his name. This is a story about Richter as an individual as well as his relationship with the world, including his professional colleagues. At least by some accounts, Richter's fame generated a certain degree of resentment among fellow scientists who saw the public acclaim more as a consequence of grandstanding than of profound scientific achievement.

Were these sentiments, which persist to the present day, fair? How did the name *Richter scale* come about? Should it have been simply the *Richter* scale, or should the names of other seismologists be attached as well? Did he properly acknowledge the contributions of colleague Beno Gutenberg? Was Richter, in the words of one later novelist, a "real SOB" who "screwed" Gutenberg out of his rightful share of fame? Such questions are difficult to answer. If it is easy to misunderstand the Richter scale, it is vastly easier to misunderstand Richter—his motivations in his interactions with the media as well as the many other facets of his enormously complicated life. Remarkably little has been written about the man, for reasons that become apparent as our story progresses. For starters, Richter was apparently not his name at birth, and therein lies the beginning of a tale of a childhood marked from the very beginning by both internal and external turbulence.

Scientists in general have a reputation for being a breed apart. It would be a magnitude 8 understatement to say that Charles Francis Richter was no exception. He was a nerd among nerds: regarded as peculiar and intensely private even by scientists' standards. And we're talking about people who put red-and-white bumper stickers on their cars that read, "If this sticker is blue, you're driving too fast."

Richter's circle of close friends and colleagues remained remarkably small throughout his life. Even his nuclear family was more nuclear than most: born into a household that included only maternal grandparents, mother, and older sister, the configuration expanded over the years only so far as to include his wife and her son from a previous marriage. Richter's wife had a sister who had two children, a son and a daughter; Richter had no children of his own, no close cousins, no nieces or nephews on his side of the family tree. His stepson never married and never had children of his own.

Richter's career had a similar nuclear quality: it began where it would eventually end, at the Caltech Seismological Laboratory in 1927, in fact before the Seismological Laboratory became part of Caltech. Hired as an assistant for a job he considered temporary, Richter never intended to become a seismologist—let alone the most famous seismologist of all time. In his mind the job represented only a brief diversion, a holding pattern in the years immediately following his completion of a Ph.D. in atomic physics. He had, not only from the outset but even decades later, every hope of some day returning to his chosen field of theoretical physics. Some biographies claim he yearned to return to astronomy, but according to what Richter wrote, astronomy had been his first scientific passion as a boy but became only a lifelong avocation from his undergraduate years onward. His formal education focused first on chemistry, and later physics.

And yet a seismologist he remained: a Seismo Lab seismologist from the start, a Seismo Lab seismologist when he retired in 1970. Few scientists in any field have careers like his, beginning and ending at a single institution. This aspect of Richter's life emerges more and more clearly as the story progresses. For now, suffice it to say that, in technical terms, Charles Richter was a homebody of nearly unprecedented proportions, even among scientists. His personal as well as his professional comfort zone, which emerges as a consequence of his extraordinarily complex and at best marginally stable personality, never stretched far beyond the boundaries of Southern California, the only home he ever really knew.

One can point to an additional key to Richter's enigmatic reputation and legacy: For a scientist of his stature, he worked with very few students or younger colleagues throughout his career. In academia one's students are one's children: they carry one's ideas, reputation, name (to some extent), and memory into both the larger world and the future. (The familial analog is widely recognized by scientists. I was once surprised and flattered to hear an eminent seismologist introduce me as a "granddaughter of sorts": my Ph.D. advisor had been one of his Ph.D. students.) Scientists tell stories about their advisors to their friends and students. Thus do oral histories—portraits of scientists as individuals as well as professionals—begin to take shape within the scientific community, if not the larger world.

Even in seismological circles Richter thus remains enigmatic. He was loathe to speak about himself, had few close colleagues his own age, few students or protégés of any sort; he spent his entire career at a single institution. Those who did know Richter are, moreover, reluctant to speak at length about the man they knew as Charlie. To some extent this reluctance bespeaks ambivalence, yet many also feel a sense of loyalty to the memory of a man they had grown to care about. Here again, the more one starts to understand the man, the more one understands the reluctance. Richter was both peculiar and private, easily hurt and famously unable to laugh at himself. The few colleagues who knew him at all well are reluctant to help paint a portrait that, viewed out of context, places undue emphasis on his abundant follies and foibles.

The few personal tidbits that are known in scientific circles tend to do just that: they suggest that Charles Francis Richter, inventor of the Richter scale, was something of a kook. He was an avid nudist; he dabbled in poetry. He sometimes showed up at work wearing two ties; when he wore only one it always sported a creative collection of stain spots. He was not in the least amused by the clever song, composed by one of his colleagues, that was performed at his retirement party. And, from some: if there were cameras around, you could count on Charlie to be there.

Every mortal has follies, every mortal has foibles; no mortal deserves to be defined by them. What is the measure of a man? Earthquakes might be difficult to measure, but they are easy to size up in comparison to the man who first measured them. To paint a full portrait—as full a portrait as is possible—of Charles Francis Richter, one must delve directly into the follies and foibles about which people have been so loathe to speak. As one makes this journey, the portrait begins to emerge at last: a man with a keen sense of humor but unable to laugh at himself; a man who never felt a strong calling to seismology but became the world's most famous seismologist; a man whose relationships with women were complicated from the day of his birth but stayed married to one woman—albeit not entirely faithfully—until her death in 1972.

The portrait of Richter includes far more than one man's fair share of tribulations: a family history rife with emotional instability, a childhood with few of the usual support systems that provide a sense of stability, personality quirks that suggest a nearly textbook set of symptoms of a profound neurological disorder, suggestions of physical ailments that would have had further deleterious effects on his sense of well-being. Many a lesser man has crumbled in the face of lesser demons than those that haunted Charles Richter's every waking hour. His difficulties did derail him for a time, nearly ending his research career before it began. Yet in the end, the work that he took on reluctantly would prove to be his salvation. It was via his work in seismology, most notably his Herculean efforts to develop the magnitude scale, that Richter was able to harness his not inconsiderable intellectual horses. Although he would never succeed in eradicating his demons, observational seismology provided such an effective outlet for his enormous drive, intellect, and talent, that he was able to make seminal contributions to the young field of seismology. In the process, he turned his name into a household world that everybody knows and almost nobody understands.

Existing brief biographies of Charles Richter, on the Web and in encyclopedias, tend to say little more than "Charles Francis Richter was born on a farm outside of Cincinnati, Ohio, in 1900. In 1935 he invented the Richter Scale." The unspoken lines that come between and after those sentences would fill a book. This is that book.

CHAPTER 2

Formative Years

There is really little in my childhood that I like
to remember or would wish to repeat.
—*Charles Richter, March 6, 1945*

EXISTING brief biographies paint a sketchy portrait of Charles Richter, the scientist and the man, and much of what they do say is misleading if not wrong. He invented the Richter scale in 1935, they will tell you, conveying a mistaken sense of the scale as a mechanical device. He was so passionately interested in earthquakes that he had a seismometer installed in his living room, the more chatty sorts of accounts say, conveying a compelling but altogether mistaken impression of a one-dimensional man.

What is written about Richter's family and childhood is sometimes wrong as well. Even when it is right, it fails to capture the essence of a singularly complicated story. *Charles Frances Richter was born on a farm in Hamilton, Ohio, on April 26, 1900.* He was, at least by his mother's later account, born Charles Frances Kinsinger: Richter was his mother's maiden name, which he adopted in early childhood but legally claimed as his own only in 1926. *His grandfather moved the family to California after his parents were divorced*, some biographies say. According to Richter his parents were married twice and divorced twice, each marriage lasting long enough (just barely, it seems) to produce a child. The initial marriage produced Margaret Rose Richter in 1892, an intriguing and accomplished individual in her own right, about whom we will hear more later. *He was raised by his grandfather after his parents divorced*, some accounts say, sometimes adding that his mother experienced episodes of emotional instability. By Richter's own account he was raised by women: his mother and older sister.

If the biographies continue—and few do—it is generally in the same vein, which is to say sometimes but not always right, and nearly always misleading. But to start this chapter of Richter's story properly, we leave existing biographies behind to begin at the beginning, relying on Richter's own account of his tumultuous childhood.

Charles Francis Kinsinger was born near Overpeck, Ohio, north of Cincinnati, on April 26, 1900. Richter's parents, Frederick William Kinsinger and Lillian Anna Richter, had married (the first time) on July 15, 1891, in Butler County, Ohio, and divorced (the first time) shortly after the birth of Margaret in 1892. Later in life Richter recalled meeting his father only once, as a boy. "He is said to have been rather erratic," Richter wrote in 1949, "full of wild ideas." Kinsinger returned to his wife in the late 1890s, just long enough to produce a second child, Charles Francis, before a second and final divorce. It is not clear how long the second marriage lasted, but by all accounts it didn't last long: the space for the father's name on Richter's birth certificate was left blank. Lillian reportedly resumed use of her maiden name, as did both children.

Although one is inclined to consider Richter the definitive authority on matters autobiographical, one must note that some of the information he knew could only have been provided to him by his mother or other older family members. Family members have, however, been known to whitewash awkward truths. Presumably Richter had no more than his mother's word that a second marriage had occurred. The conspicuous blank on the birth certificate suggests that the birth could in fact have been illegitimate. For that matter, viewing available evidence with dispassion, one notes that we have only hearsay evidence that Kinsinger was indeed Richter's father.

Richter may not have ever known that his (purported) father was one of five children born to Joseph Kinsinger and Helen Kennel, the second child and first boy. Several of his siblings had children of their own: Richter apparently never knew that he had several first cousins, almost all of whom remained in Ohio. Frederick moreover remarried in 1909, a woman by name of Olive G. Burrnesker. There is no indication that Richter was ever aware of, let alone met, the midwestern branch of his family tree. He did know a few bits of information about his father's side of the family: that they were Amish, in particular Mennonite, and had been in the United States for about a hundred years. One suspects the twice-divorced Frederick Kinsinger must have been something of a black sheep among his especially pious flock. One also wonders if his heritage didn't leave its mark on his son's psyche: any isolated and limited gene pool can start to produce more than its share of offspring with unfortunate genetic traits. (That this statement may offend does not, unfortunately, negate its veracity. In Geague County, Ohio, for example, the Amish represent about 10 percent of the population but fully half of the special-needs children in local schools, an imbalance that cannot be explained by the relative number of children in the Amish versus the overall population.)

As a physical specimen Charles was scarcely well assembled. A bout of infantile cholera at the age of fifteen months nearly ended his life before it had really even begun. He remained small and weak throughout his childhood. At thirteen, he was, by his own later account, five feet, eight inches tall and a scant seventy-seven pounds—impossibly skinny, it seems. Yet one would expect that the man who first measured earthquakes would know from whence he spoke regarding measurements of himself. He would never grow much beyond his height at fifteen.

Richter's mother educated him at home during his early childhood years: scarcely unusual in rural Ohio in the early years of the twentieth century. Lillian Anna Richter had been a schoolteacher before her own children were born. Among the more poignant items among the Richter Papers at the Caltech archives one finds a German child's game, "Was kann ich brauchen?" contributed by Richter's longtime friend Jerene Hewett, who notes that Richter's mother would have almost surely used the game in her teachings. A scribbled price of two dollars still graces the lid of the box. (In 1900 the average American family had an income of three thousand dollars, compared to over thirty thousand dollars today: a two-dollar game was no cheap plaything.)

Richter's maternal grandfather, Charles Otto Richter, retired in the early 1900s from his position with a large manufacturing firm, Hooven-Owens-Rentschler Company. Soon thereafter he sold the family house and farm and moved the family to Los Angeles. Such a move would have been considerably less commonplace in 1909 than it is today. By Richter's account, his grandfather had decided on a change of scenery after hearing glowing accounts of the frontier state of California. The death of his wife, Richter's grandmother, in 1907 may have also inspired the patriarch to pull up stakes.

The family moved as a foursome: Richter, his grandfather and mother, and older sister. They first lived in a small house in the Wilshire district, near the corner of Eighth Avenue and Kingsley Drive. Few original single-family homes have survived to the present day in this neighborhood, where streets are now lined with apartment buildings and businesses in what is now known as Koreatown. A few wood-shingled homes of relatively modest construction, now converted to businesses, can be found near the intersection. In 1925 the family moved to a house at 723 South Bronson Avenue, a little over a mile west of their first house. The family, in particular Richter's mother and sister, would remain in this house, along with an assortment of cats, for many years. Richter himself remained in this home until he moved to Pasadena with his wife in 1936. A niece of Richter's wife, Dorothy Crouse, recalled childhood visits to a grand house: two full stories, an elegant chandelier and velvet drapes in the informal eating area, a stately grandfather clock on the landing of the staircase, enough bedrooms for the family as well as a formal guest room. This house, which had been built in 1911, still stands: a large and gracious two-story Victorian. By no means extravagant by modern standards, one has to remember that the average American house size was only about 1,200 square feet in 1940, and did not reach 1,500 square feet until 1960. In the eyes of a young woman who had grown up in a far more typical home of its day, the house was nothing less than a mansion. Crouse's starry-eyed impressions were only enhanced when she made friends with a playmate across the street, a girl who had once been in a movie with Shirley Temple. Crouse was always under the impression that the purchase of the home, as well as the family's lifestyle, had been underwritten by family money—perhaps the sale of Richter's grandfather's farm and home in Ohio.

There was turbulence beneath the waters of Richter's early family life, its elegant trappings notwithstanding. "I am convinced," he wrote, "that it was my mother's excessive and persistent attachment to her father that distorted her life—and to some extent, mine." His parents had "made the mistake of trying to live with her parents" after their (first) marriage. "Subsequently my father brought suit for alienation of affections." No elaboration is offered beyond these words, no details to explain fully the nature of the "excessive and persistent attachment." Lillian Richter's relationship with her father might have been no more than that of a run-of-the-mill daddy's girl; however, the range of possibilities clearly runs the gamut. Some secrets remain unspoken during even a man's (or woman's) most candid moments.

Some biographies of Richter say that because of his mother's instability, there were periods when he was essentially raised by his grandfather. The ultimate wellspring of this information remains unclear. By Richter's own account, by the time he reached an age in childhood when memories are preserved clearly, his grandfather had become the "old man of the house." A tendency towards early dementia—possibly early-onset Alzheimer's, perhaps a serious metabolic disorder—plagued his mother's side of the family, including the grandfather. Whatever he had been earlier in life, time and the ravages of imperfect genes clearly muted the grandfather's

Fig. 2.1. House on Bronson Avenue in Los Angeles where the Ritcher family resided between 1935 and the mid-1950s. (Photo by author.)

personality in his later years—leaving Richter in the hands of the first two women in his life, his mother and a sister who "rather bossed me around."

California did not welcome young Charles with open arms. Richter described his early childhood in Ohio as having been pleasant, adding that it "had been a shock to encounter city life and city children after being alone so much." As one might imagine, the social web of a small-town society (which Los Angeles was in 1909) did not rush to embrace a divorcee with two young children. One might further imagine their reception to have been exacerbated by the peculiar nature of this particular "broken home." Charles did attend the local public school, Hobart Boulevard School, which can still be found—although not in its original incarnation—at 980 S. Hobart Street in what is now the Wilshire district. Between an awkward family situation, his frail health and introverted nature, and his prior lack of formal schooling, his initial foray into formal education did not go well. A fourth-grade report card reveals an F in arithmetic. (At the time F stood for "fair," rather than the "fail" it later represented. Still, it is scarcely the grade in arithmetic one would expect of the future inventor of the Richter scale.)

Richter paints a self-portrait of a poignantly lonely child. Perhaps one can understand why his attentions were drawn upwards, towards the nighttime sky: what would remain a lifelong fascination with astronomy and, later, science fiction. Among the collection of Richter's papers at the Caltech archives one finds a box full of classic science fiction magazines: *Amazing Stories, Thrilling Wonder Stories, Other Worlds*, and more, spanning the years 1926–55.

By the age of fifteen Richter had gotten involved with serious amateur astronomy, making regular and detailed observations of so-called variable stars. As their name suggests, such stars vary in their light output, the consequence of either internal changes or the eclipsing of the star's light by another body. Some variable stars pulse, others—such as supernovas—literally explode dramatically in a final paroxysm marking the star's demise. Even amateur observations of variable stars can be valuable to astronomers. The American Association of Variable Star Observers was founded in 1911 and remains active to this day. Richter sent a series of observations to the AAVSO, members of which apparently valued his contributions: when he stopped sending reports in 1917, a representative of the association wrote to Richter to let him know that his reports were missed.

Nowhere in Richter's writings does he speak of a childhood interest in the planet beneath his feet, or in earthquakes specifically, although he did later recall feeling an earthquake in 1910. This would have been a temblor that struck south of Riverside on May 15, now estimated to have been of magnitude 5.5. With a location some fifty miles south-southeast of Los Angeles, Richter would have most likely felt only gentle morning (7:47 a.m.) rumblings. Several decades later this earthquake would merit two lines, under "Other Shocks of Interest," in Richter's landmark 1958 textbook, *Elementary Seismology*. If earthquakes did not capture young Charles's attentions, it was possibly because there were very few earthquakes to capture anyone's attention in Southern California during the first two decades of the twentieth century. Moreover, while Charles was occupied with the business of growing up, prevailing business interests were occupied with downplaying earthquakes and earthquake hazard in Southern California. The earth itself cooperated for a time, producing few earthquakes of consequence in the opening decades of the twentieth century. (A magnitude 4.9 temblor centered in Inglewood, south of Los Angeles, did strike on June 21, 1920, causing some damage to vulnerable structures. At this time, however, Richter was at Stanford.)

In 1912 Richter's mother enrolled him in a preparatory school associated with the University of Southern California near downtown Los Angeles. This institution would have been superior to the local public schools at which Charles made his awkward educational debut. Schoolwork from his high school years reveals far more satisfactory marks: scores of 92–100 percent on papers in subjects ranging from math (he seemed to especially excel in geometry) to economics, English, and Latin. He talked later of being drawn naturally to the sciences by children's books around the house; these books had first drawn his interest to the stars. His studies appear to have progressed satisfactorily in other areas as well. One is, of course, unlikely to save the papers with poor marks, but Richter graduated from high school in 1916, at the age of sixteen. His high marks in Latin suggest that by his teen years he was well on his way to becoming a linguist: by his adult years he had a good reading knowledge of German, French, Italian, Spanish, Portuguese, and Russian, and could converse fluently in many of these languages.

Richter also developed a lifelong passion for books. Later in life the walls of his homes were lined with bookcases filled with thousands of titles covering a wide range of topics in addition to the science tomes one would have expected to find. The collection reflected Richter's lifelong interest in both the stars and science fiction. Near the end of his life in 1984 he donated his science fiction collection to the Caltech library: 275 hardcover books, 1,200 paperbacks, and 1,450 classic science fiction magazines. Some of the paperbacks were distributed among houses that Caltech uses for overflow student housing: later generations of students would be surprised to leaf through a book and find it signed by the most famous seismologist of all time. (Richter apparently scrawled his signature on the first page of every book he ever owned.)

By Richter's own account his social adjustment did not improve during his teen years. He had, however, by this time discovered not only the pleasures of amateur stargazing and of reading scientific books; he had also discovered the local mountains, initially in 1916 via hikes organized by the Lorquin Club. In 1916 the Lorquin Natural History Club, later the Lorquin Entomological Society, comprised only a dozen or so members who went hiking and backpacking in the mountains to look at and collect local plants. At the time, botany ranked high among Richter's scientific interests. For the first time he discovered the setting that would remain his sanctuary throughout his life. "The out of doors," he wrote in 1945, "the walked trail, the climbed mountains—these were my real emotional roots before the burst of adolescence."

Richter enrolled as a freshman at USC—a natural next step—but remained there only a year before moving to finish his undergraduate education at Stanford. Of this move he later wrote that his sister was at Stanford, having graduated from that institution earlier with a degree in 1918 (later to earn a Ph.D. there in 1927), and he found it to be an attractive situation. He wrote that "Stanford is a state of mind, and I don't think one can sum it up very easily in a few words." He began a course in chemistry, but "that didn't seem to be a satisfactory adjustment. Gradually I got into physics, which was more congenial." In a 1979 interview he elaborated: "at that time I was quite nervous and tended not to be neat, particularly with my hands, and this is fatal in a chemistry laboratory. So after some unfortunate experiences, I felt that this wasn't for me." One is inclined to suspect that his instructors felt the same way. However, a report card from 1919 reveals an A in Principles of Chemistry and only a C in Physics, as well as an A in applied math and a B in elementary French. In the long run physics nonetheless proved to be a much better fit. He completed a bachelor's degree in physics in 1920 and embarked immediately thereafter on a path to a graduate degree in the same subject, also at Stanford.

Notwithstanding the timely progression of his matriculation, the tumult of his early childhood continued to ferment, in short order developing into full-fledged storms. In 1919 he exchanged correspondence with a former high school teacher, Howard Leslie Hunt. Hunt counseled the young man to look for guidance in the Scriptures; this would not prove to be useful advice to a young man who remained at best agnostic throughout his life. Richter's inner turmoil worsened. In one especially terse account Richter described his state as "Adolescent confusion and nervous near-breakdown, 1921." A year after beginning his graduate studies Richter returned home to Los Angeles, "bag and baggage," where "my mother had the good sense to refer me to a psychiatrist." His mother dispatched him to the care of Dr. Ross Moore, who had previously treated Richter's sister.

Richter's "adolescent confusions" were severe enough to land him in a private sanitarium under Dr. Moore's care, where he remained, "with interruptions," for about a year. The treatment did help Richter regain his emotional footing. In 1922 he took his first employment as a messenger boy for the Los Angeles County Museum. From there he went on to work as a clerk for a hardware company, allowing him to save some money and eventually return to academic work at the newly established California Institute of Technology in Pasadena.

At this juncture one can leave the strictly historical for a brief venture into the more suppositional. By sixteen Richter had revealed himself to be

Fig. 2.2. Charles Ritcher, circa. 1925. (Photo courtesy of Caltech Seismological Laboratory.)

both talented and troubled. The latter is no wonder. As we have already seen, his family life was marked from the beginning by instability; his childhood was turbulent and missing the usual social trappings that help guide any young person to adulthood. If it takes a village to raise a child, Richter never had a chance: he never had a village. His world revolved around his immediate family, an insular and dysfunctional family at that. The transition from high school to college represents a leap of faith and courage for any teenager, no matter how healthy and well adjusted. Leaving the nest is hard. Leaving the nest with the baggage that Richter carried as a young man more than hard—perhaps the only surprise is that his emotional breakdown did not happen sooner.

Richter is, nonetheless, hardly the only young man or woman to have left the nest and arrived at a state of emotional crisis on the brink of adulthood. The move back to Los Angeles—back home—undoubtedly went a long way in restoring some semblance of order in young Charles's life. As his story unfolds, one theme becomes clear: Richter's emotional comfort zone never extended beyond the confines of the only hometown he had really known. Yet even within the confines of home, "comfortable" does not emerge as a particularly apt word to describe Richter as a young man and budding scientist. Even as he regained a fragile toehold in the outside world, he kept a personal journal, dated from June 1926 to June 23, 1928, while he was a graduate student at Caltech. This remarkable document reveals with stunning clarity (albeit atrocious handwriting) the degree of inner turmoil that continued to plague him during his graduate studies.

Much of Richter's graduate school diary chronicles events as they unfolded, revealing his state of mind throughout these years. Midway through his journal, however, he looks back at the time leading up to the "mental near-breakdown" that ended his studies at Stanford and provides a measure of illumination about these turbulent years. In handwriting that disintegrates from bad to worse, Richter recounts an encounter with a man whose name he cannot recall exactly, but believes was something like Frank Reed. The encounter took place at a time when he was discouraged about his performance at Stanford. "This must have been due," he wrote, "to getting out of high school into the university freshman year, and finding myself suddenly much less important." Such a feeling is scarcely unique among freshmen at elite universities who suddenly find themselves just one of many enormously talented young people. Yet consider what Richter goes on to say: "At Stanford matters were worse; I felt still more out of place." Clearly, unlike many better-adjusted freshman, feeling out of place was nothing new for Charles Richter.

Richter describes himself as sitting in his room, "moping," towards the end of the year, and encountering Frank Reed. Richter was not well acquainted with Reed, but had heard Margaret speak of him. Reed was "one of the older men, and a thoroughly fine fellow—clean, straightforward, a real 'Stanford Gent.' " Richter found himself expressing his disappointments to Reed, who responded with nothing but reassurance: "He encouraged me, talked to me quite kindly—said he knew my sister—and got me somewhat out of the dumps (though not definitely)."

Richter then writes, "Now this evening, as I called this to mind, I suddenly burst into tears. I was actually astonished. I could not understand what was stirring me deeply; yet every time I returned to the recollection there was the same burst of weeping." In a crushing realization that came only years after the chance, casual encounter, Richter understood that Reed represented "exactly the sort of man" he himself wanted to be, the "finest kind of Stanford man." He had come to Stanford with the "hope of associating with such men, and learning to be like them," but most of them had left for World War I (this was 1921), and, Richter adds, "I poked off into a corner." Richter in fact enlisted in the Students' Army Training Course (SATC) at Stanford, and served precisely two months, eleven days, before receiving an honorable discharge. He never elaborated on the reasons behind the premature end to his military career: presumably the poorly coordinated Charles Richter and the highly regimented U.S. Army discovered they were not right for one another.

Returning to Frank Reed, Richter went on to say, "Part of the appeal of such a personality to me was that I knew that such characters have strong sex appeal to my sort of women—my sister and mother, for instance."

"This, then, is my secret," Richter wrote, "and at this time point I burst out into the most awful flood of tears I have ever known. For all these years I have been carry[ing] about with me an ideal of manhood of gentlemanly living; an ideal to which I had every reason to think I might attain; an ideal which was not altogether imaginary, but was at least based on real personalities. I have miserably failed to measure up to such an ideal. Therefore I have been disappointed, disgusted, at odds with myself. No wonder I have had a inferiority complex!"

Richter's reflections took him to the next obvious question: "Where did this ideal originate?" He answers his own question, pointing to a chemistry instructor during his second year of high school, back when Charles was an awkward and undersized boy of fifteen. The instructor, Mr. Wheeler, was a tall, well-built man with a forceful personality. Wheeler caught the attention of Richter's mother, who was "much pleased with his appearance." "I remember remarking," Richter wrote, "that in the course of time I might develop into just such a big man."

Fifteen-year-old boys are, of course, known to set their sights on impossible ideals—professional athletes, rock stars, movie stars—than a well-built chemistry instructor with a commanding presence. And yet, as Richter realized, the failure to live up to this ideal was all the more crushing because it was attainable. In due course most older teens will come to understand that they will never measure up to the idol of their younger teen years: that they are no Michael Jordan or Mia Hamm. By the end of their second decade most teens understand that most people are not remarkable physical specimens. Richter, however, had not measured himself against a star athlete but rather a teacher—the sort of person a child might plausibly grow into one day.

As it turned out, even Richter's modest ideal remained well beyond the reach of a painfully introverted, awkward, physically weak, and undersized young man. Moreover, his later writings make clear that his ideal man was not only one who fit a certain mold, but also one who made a strong impression on Richter's mother. He yearned from a young age to be a certain type of man—a type of man that he was not, and was not likely to ever be; the type of man to whom his mother was attracted.

Freud would have had a field day with this. Indeed, Lillian Anna Richter emerges as an enigmatic character in the drama that was Charles Richter's life. We have no evidence that she fell short substantively in her commitment to her children. A schoolteacher by training, she taught them at home during their early childhoods, sought out appropriate educational opportunities, and, when necessary, appropriate psychiatric care. Richter's niece by marriage later recalled "Mrs. Richter" warmly, a grandmotherly figure who had been happy to read bedtime stories to visiting youngsters. And yet, by Richter's account, his mother's life had been "distorted" by an excessive attachment to her own father; there were further indications that she remained drawn throughout life to unattainable idealizations that clearly left an imprint on her only son.

We are, in the end, left with a vexingly incomplete Lillian Anna Richter, a woman who was not famous enough to be written about in her own right and too loved, it seems, to be written about at any length by her offspring, both of whom achieved a measure of fame. Basic biographical information reveals that she was born in January 1871 in the state of New York and was married (the first time) in Butler County, Ohio, in 1891. Her children were born in 1892 and 1900. She died in October of 1953 at the age of eighty-one. (The Richter genes, however flawed, do appear to have been programmed for longevity.)

Lillian Richter makes remarkably few appearances in her son's voluminous writings, including detailed journal entries and drafts of letters. One would perhaps not exchange many letters with one's mother if proximity permitted direct communication during much of one's life. Such was the case for Richter, who for most of his life was no farther away from his mother than Pasadena is away from Los Angeles (roughly eleven miles).

Lillian's absence from Richter's journal pages does perhaps hint that some mother-son issues were left unspoken. Among his papers one finds many drafts of letters to other people throughout his life, but none to his mother can be found among his personal correspondence files. One draft of a letter among Richter's papers is, however, hidden among an otherwise disorganized collection of writing, mostly poetry—it looks like a scrap of paper that survived by accident rather than intent. This draft is not addressed to anyone but can only have been a letter from son to mother during Richter's years at Stanford. It begins, "Now please don't tell me to stay at home and work around there (in that mess of a garage, for example!). I must get out among *men*." Richter goes on to say, "I have been—forgive me, mother-tied to your apron strings all these years, until I have become almost effeminate. At the University, I've been as dependent on you as if I were at home. The only exception was during the S.A.T.C. If that had lasted a bit longer, it might have made a man out of me."

The letter continues: "I think you understand by now that I don't want any office job; neither do I want any job connected with my science—I want to forget that for a while. First I want to make inquires about a possible job in Yosemite—they are open all year now, failing that, one of the all year jobs near Los Angeles." The draft ends abruptly, as if it wasn't completed or was once part of a longer letter.

The single page nevertheless speaks volumes: compelling evidence that Richter did not have an altogether happy relationship with the first woman in his life. At the same time, Richter was scarcely the first adolescent to lash out in harsh terms towards a parent from whom they must establish independence. We will never know if Richter's difficulties with his mother were more than garden-variety sorts of adolescent tribulations. One does tend to suspect something more, if only because of evidence that both mother and son were strong and complex personalities in their own rights—not to mention the fact that Richter had been raised in what was in its day a most unusual single-parent household. In the end, however, our portrait of Lillian Anna Richter, the woman who headed that household and had a profound influence on her only son, remains fragmentary. One is inclined to think that Lillian probably did struggle with demons of her own. Yet of all the painful secrets that Richter held dear throughout his life, few seem to have been held as dearly.

The novelist could fill out the portrait of Lillian Richter. One can imagine her in a number of costumes: some degree of Electra complex, clearly (*but how much*?); perhaps a personality disorder. The novelist could paint the picture; the biographer can only acknowledge a picture that isn't there—or at least, is not altogether complete. Whatever the true nature of the portrait, this much seems certain: Lillian Richter was drawn to idealizations of men rather than realizations of men, and young Charles set his own ideals accordingly.

One can scarcely imagine a surer recipe for emotional disaster for a boy: his own father absent, his grandfather a shadow of his former self by the time Richter reached preadolescence—no real man of the house, the woman of the house attracted by unrealistic ideals, the son not remotely close to those ideals as a physical or emotional specimen. Had Richter been a different sort of individual—a more fundamentally sturdy sort—he might have emerged from this upbringing with no more than any young man's fair share of neuroses. As we will see, however, he was in all likelihood anything but neurologically sturdy. Profoundly introverted, shy, and socially awkward, but not only this: the real roots of Charles Richter's inner turmoil very likely ran far deeper.

Richter's graduate school journal reveals the depth and breadth of the demons that for a time got the best of him. A picture begins to emerge of a man who would have struggled to cope with the world even under the best of family circumstances; a man whose brain, while extraordinarily nimble, was also extraordinarily wired.

In modern times educators and parents are attuned to children with special needs. Learning disabilities, so-called mood disorders: such tribulations are today recognized, labeled, treated, and accommodated. It was not always so. Ritalin—the first common medical treatment for attention deficit disorder—was invented only in 1980. Just a few decades ago a forgetful, scatterbrained child was likely to be seen, in the eyes of parents and teachers, as a lazy or stupid child. Not many decades ago, a child with a more unruly sort of disorder, such as attention deficit hyperactivity disorder or bipolar disorder, might been classified as a juvenile delinquent. In the past, a child with limited proficiency in reading social cues might have been perceived as any number of things, none of them kind: geek, nerd, loner, loser . . . or the equivalent in the parlance of the day.

The full picture of Richter emerges only later, after one considers the follies and foibles—as well as the struggles and successes—of the man through his adult years. Returning to Charles as a twenty-one-year-old Stanford student, one finds a portrait of a young man who was, in short, a mess. Prodigious academic talents carried him through Stanford University with the degree in physics; equally prodigious emotional turmoil led to a breakdown severe enough to land him in a sanitarium.

Richter's psychiatrist, Dr. Ross Moore, makes only brief cameo appearances in Richter's journals and letters. All indications, however, suggest his treatments to have been effective. After a year under Dr. Moore's care, Richter made his first tentative steps back into the outside world, first securing gainful employment within sheltered waters but soon moving back to the world of research science. Moore also encouraged Richter to try his hand at expression in verse, what would become a critical outlet for his emotional energies throughout his life. As late as 1949 Richter would acknowledge his therapist's continued care with appreciation. It was probably beyond Dr. Moore's ability—or any psychiatrist—to "fix" all of Richter's problems. Without question his childhood left scars, as evidenced by the epigraph to this chapter, Richter's words at age forty-five: "There is really little in my childhood that I like to remember or would wish to repeat." These words were part of a single-page, typewritten document, addressed to no one, apparently sent to no one, beginning with, "Occasion once more to take stock of myself and write out ideas and plans. The direct inspiration of last summer has long been spent; its effect remains, and I believe that without much effort I could turn again to poetic expression." Exactly what prompted this occasion, and the nature of the inspiration from the summer of 1944, he does not say, although a later chapter will perhaps provide hints.

In any case, a poem penned later, in January 1967, reveals an even more poignant sense of anguish about his formative years:

> Contentment, so it seems, cannot be trusted;
> Not only are there perils in the world,
> Lightning and falling stones and open pits,
> But underneath my mind the unschooled ape,
> The grasping infant and the loveless boy,
> Still live in shadow waiting for their turns.
> They do not disturb me much nowadays,

> Yielding to mingled common sense and laughter;
> But yet fatigue, or any sudden shock,
> Or even joy—God help me, even a kiss—
> May stir these creatures from their hiding places.

The creatures from Charles Richter's childhood were, it seems, never far from his heart, well into and beyond middle age.

One is nonetheless inclined to view not only Dr. Moore but also Richter himself with respect. No fairy godmother waved a magic wand to banish the storms that battered the young man's soul: they remained an integral part of him throughout his life. Rather, Richter turned his relentless energies and intellect towards the task of understanding his own failings and, further, of coming to grips with them. This, then brings us to the next chapter in Richter's life: his graduate school tenure at Caltech—six remarkable years during which he attained a mastery of modern atomic physics and, perhaps more impressively, began to attain at least some mastery of himself. Before we move forward to this time, however, we pause to meet another key supporting character in Richter's life: his only sibling, Margaret Rose Richter.

Margaret Rose

> The poet's urge, a knotted cord,
> Is laid upon my shrinking flesh,
>
> Exacting payment, red and fresh,
> For every line, for every word.
>
> —*Margaret Rose Richter, "Flagellant," 1933*

Richter's complicated relationships with women began the day he was born. We have hints of the difficulties between mother and son, although in the end we are left with less than a full portrait of Lillian Anna Richter. Of Richter's only sister, Margaret Rose Richter, we know a good deal more, not only because of the fame she achieved in her own right but also by virtue of the words her brother left behind, and the words of hers that he kept. Yet Richter's writings represent a view of her life observed through a filter marred by its own distortions and from a proximity that rendered objectivity impossible. In his writings about Margaret, Charles had not only the lack of perspective that any human being has their only sibling; his understanding of her problems and their enormously complex relationship was further influenced by difficulties of his own. Some of what Richter writes is easy to understand; other snippets lead to far more questions than answers.

In at least some respects the early brother-sister relationship fit the expected pattern for siblings separated by a large number of years growing up in a dysfunctional household. Brother and sister might have been drawn together more closely than some by virtue of their mother's own emotional difficulties and their unconventional home life. By Richter's account Margaret "rather bossed me around"—one can scarcely expect it to have been otherwise for a young boy in, effectively, a single-parent household with a much older sister. Margaret was, moreover, a strong, talented, and complex individual in her own right. She graduated from the Los Angeles Polytechnic

High School, the school of the Los Angeles Public Library. The fact that she received a Ph.D. as well as a bachelor's degree from Stanford University in the early decades of the twentieth century speaks volumes by itself. While Stanford accepted female students in its early days, an amendment to its founding grant in 1899 limited the number of female students to five hundred. Only in 1933 would a resolution increase this number to one thousand. At the time that Margaret received her bachelor's degree, only a scant 1 to 2 percent of young American women were similarly well educated.

Margaret had returned to Los Angeles and was living with her mother when Charles and his wife Lillian (nee Brand) were married on July 19, 1928. By Richter's own account, he "made the mistake of taking Lillian to my mother's house to live. Lillian was a little wild and erratic at that time; mother did not understand her, and my sister made matters worse with constant hostile suggestions." Later chapters will illuminate further the nature of Lillian's early—and later—wild ways.

Returning to the story at hand, Richter's marriage appears to have disrupted his formerly nuclear family household. Not long afterward, Margaret secured a teaching position at the University of Arkansas. Richter's mother moved with her daughter, presumably to keep an eye on her more troubled child, but perhaps also to leave the house on Bronson Avenue to the newlyweds. By all indications Margaret was indeed troubled: we know that she was under the care of psychiatrist Dr. Ross Moore before her brother was entrusted to his capable hands in 1921. The nine-year gap between her undergraduate and graduate degrees suggests that her studies, like those of her brother, had been interrupted.

By all indications, however, Margaret also possessed considerable talents in her own right. She wrote poetry successfully enough to win prizes and have her work published in magazines and anthologies. Among Richter's papers at the Caltech archives one finds a bound book, over an inch thick, full of clippings of Margaret's published poems, articles about her, and a poetry column that she wrote for publications such as *Poetry World* and *The Poet.*

One early poem, dated 1933, describes with clarity a poet's soul:

FLAGELLANT

The poet's urge, a knotted cord,
Is laid upon my shrinking flesh,

Exacting payment, red and fresh,
For every line, for every word.

On up the path I stumble, pant,
Gasping for more ecstatic pain,
And swing the lash again, again;
I walk an eager flagellant.

O scourge of beauty, scourge of flame,
For every sting of anguish blest,
Curl closer round my naked breast;
Wound me; I bleed in beauty's name!

Another poem describes the life choices made by a woman at a time when serious career ambitions and a family life were mutually exclusive:

TRAVAIL

They hungered for a mate and motherhood,
And they have chosen children of the flesh.
For offspring of their bodies and their blood
They suffer Eve's old agony afresh.
That in their children they may live again,
That they may have companions for a while,
They make themselves a sacrifice to pain—
And look upon the long years with a smile.
But I have chosen children of the mind,
Knowing I shall have only thoughts to fill
The heavy years I slowly leave behind,
The lonely days that will grow lonelier still.
Then let them not account my travail less,
Who have not known my spirit's weariness.

Indeed, Margaret would never marry or have children. In *A Room of One's Own*," her 1929 study of the conditions under which women write, Virginia Woolf imagined that Shakespeare had a sister, Judith, who shared her brother's writing prowess. Woolf paints a convincing portrait: a brilliant woman at a time when women had no options, her genius stymied, pushed inexorably to take her own life on a cold winter's night. Born several centuries later than Elizabethan times, Margaret Rose Richter had options, and she did not kill herself on a winter night or any other night. Yet she also knew, and expressed with eloquence in "Travail," that a career and family life were either-or options for women of the early twentieth century.

Her elegant verse notwithstanding, the demons that tormented Margaret were less benign than those her brother struggled with; perhaps society was less tolerant of them as well. Richter wrote, "When Margaret finally became, to put it mildly, very neurotic, had emotional outbreaks with anger and weeping in her classes, etc., she was finally forced to leave Arkansas." This would have been around 1935. "She returned with the fixed idea that she had been the victim of a conspiracy. (Actually she had had her cap set for a man who, I suspect, was not really interested at all; she was sure others had deliberately broken it up.) After her return from Arkansas, people in Los Angeles were, she insisted, always doing things to her; there were 'annoting' [*sic*] and 'mysterious' telephone calls, so that there finally was an unlisted number." This and other accounts point towards paranoia: a degree of suspiciousness or mistrust that is highly exaggerated if not completely unwarranted.

Well known as half of a complex combined disorder, paranoia can manifest itself as part of paranoid schizophrenia, a condition with profound and frequently disastrous consequences. Paranoid schizophrenics may hear voices and experience bizarre delusions or hallucinations. They are not uncommonly rendered incapable of functioning in society, although medications can now control the symptoms, to some extent. Nobel Prize–winning mathematician John Nash suffered from this disease. Among those afflicted with the disorder, Nash was distinguished not only by the brilliance of his accomplishments before he was afflicted but also by the fact that he was able to make his way back from darkest reaches of the illness.

Paranoia, however, can be present by itself. Individuals with more mild paranoid disorders may be inclined towards delusions but are often able to maintain gainful employment and a place, albeit perhaps not an entirely comfortable one, in society. Some scientists believe that, while paranoid traits are probably innate, the disorder can be heightened by high stress levels. There is, for example, evidence that paranoia is more prevalent among immigrants and prisoners of war. Individuals thrust into new and highly stressful situations sometimes develop "acute paranoia," a form of the disorder in which especially intense delusions develop over, typically, a short time period.

It does not require a great stretch of imagination to imagine that, like her brother, Margaret had a smaller emotional comfort zone than most. Like him, she completed a degree at an institution several hundred miles from home, but also like him, she thereafter returned to home—and to psychiatric care. (Stanford University was also, one notes, founded specifically with the intention of providing a sheltered, genteel atmosphere on "the Farm," in which young people could devote their full attentions to their studies with a minimum of worldly distraction.) Margaret clearly brought emotional baggage with her when she moved to Arkansas, but the move itself might well have exacerbated her problems.

Margaret moved back to California in 1935 or 1936, an occasion that inspired Charles and Lillian to leave the family homestead once and for all. Margaret remained in California for only a few years, during which time she offered poetry workshops in the family home on South Bronson Avenue and continued to write verse. In 1937 she penned one lovely poem that seems to belie her inner turmoil:

> GHOSTS
>
> These are the only ghosts
> That ever haunted me:
> The ghost of whitening breakers
> Upon a cobalt sea;
>
> The ghost of ice-barred ranges,
> Reaching toward the moon;
> The ghost of melting snow drifts,
> Through mountain meadows strewn;
>
> The ghost of a cataract,
> Rock-shattered to a dream;
> The ghost of white azaleas,
> Their fingers in the stream
>
> These are the only ghosts
> That ever haunted me:
> The hovering wraiths of beauty
> That will not set me free.

By the end of the 1930s Margaret had secured a teaching position at Columbia University in New York, but her emotional difficulties soon resurfaced. Among Richter's papers one finds a letter from William McCastline, M.D., dated March 19, 1940, informing Richter that "situations that in the past made difficulties for [her] at Stanford and the University of Arkansas are still present." The doctor wrote of an unspecified "unpleasant occurrence" that compelled another young woman in the residence hall to leave her own room; yet Margaret refused to consider living elsewhere because, in the doctor's words, she felt that the "dormitory protects her from her enemies."

Dr. McCastline urged Richter to encourage his sister to move back to California, but in his reply, Richter told the doctor—one suspects astutely—that such action would only enrage his sister and convince her that he and the doctor were conspiring against her. Richter instead urged the doctor to contact Dr. Ross Moore. The ultimate outcome of this exchange remains unclear: however, Margaret did indeed return to Los Angeles in the early 1940s.

Margaret's scrapbook reveals a steady output of poetry through 1941; after this date the thinning pages suggest a decline in faculties. Between June 1941 and May 1946 the pages include few poems but are filled instead with a poetry column that she wrote for *The Matrix*, the publication of Theta Sigma Phi, a national professional journalism fraternity. The columns petered out by 1946, a final one appearing in the April—May issue that year. The following pages of the scrapbook are mostly newspaper clippings: little articles and, finally, fashion photographs. By 1949 Richter observed that his sister, "who [had] always been neurotic, shows definite deterioration at her present age," then fifty-seven. It appears to have been Margaret's decline that prompted Richter to write a soul-baring letter to Dr. Moriarty in 1949—a letter that provides key and detailed biographical information found nowhere else in his papers. Dr. Moriarty was a specialist in psychiatry and neurology whom Richter had apparently consulted in addition to the routine medical care he received from his regular physician.

Richter wrote in his letter about a familial tendency toward "insanity due to arteriosclerosis"; to modern ears his accounts suggest early-onset Alzheimer's, which today is known to have a significant genetic component. Fewer than 10 percent of Alzheimer's cases are diagnosed in individuals younger than sixty-five, the cut-off age used to separate "early onset" from the more typical development of the disease. Alzheimer's can in fact strike as early as a person's thirties. One thus starts to wonder if disease might have been behind Margaret's paranoid tendencies and outbursts earlier in her life.

Or perhaps the culprit was something other than Alzheimer's. Paranoia can be a feature of a number of mental illnesses, including depression and dementia. It can also result from infections such as meningitis and sepsis, as well as from severe metabolic disorders such as hypoglycemia and vitamin B-12 deficiency. The link between these disorders and apparent neurotic behavior remains on the fringes of the mainstream medical community, but reputable medical experts point to studies suggesting the link is real. Vitamin B-12 is vital in red blood cell formation; its deficiency can lead to a serious condition known as pernicious anemia. Among the symptoms of this condition are mood swings, paranoia, irritability, confusion, dementia, hallucinations, and mania.

Hypoglycemia—a once trendy diagnosis but a very real condition nonetheless—can also lead to symptoms that mimic neurologic or psychiatric disorders, including bipolar disorder and hysteria. Hypoglycemics can experience enormous variations in blood sugar levels; frequently the blood sugar level is far too low. When blood sugar drops rapidly, symptoms such as weakness, nervousness, irritability, and mental dullness ("brain fog") can occur. According to some experts a second category of symptoms can arise when stress causes the so-called fight-or-flight response, whereby the body secretes the hormone adrenalin. The combination of low blood sugar and high adrenalin can further affect mood, causing shaking, trembling, anxiety, irritability, and paranoia. Moreover, the usual soothing effects of the body's natural endorphins are largely negated in this situation. The individual finds him or herself not only terribly anxious but also lacking the normal physiological mechanisms that self-regulate mood.

Some medical professionals are convinced of a much stronger dietary component to (apparent) mental illness than is recognized by the mainstream medical community. Certainly, as many dieters have learned in recent years, a diet low in simple, refined sugars—a large component of the prescribed dietary treatment for hypoglycemia—can raise energy levels and improve mood, as well as to improve the body's overall regulatory systems.

Viewing any set of symptoms retrospectively, one can obviously never pinpoint the neurobiological disorder responsible for Margaret Richter's emotional problems—if indeed one was responsible. Quite plausibly more than one factor contributed. One further clue, however, emerges when one considers her activities during the later years of her life. Margaret's condition deteriorated until she was declared incompetent in late 1952, after which time she spent several years in a state mental institution in Sacramento. Yet in 1960 she was released, and in 1962 she was lucid enough to be writing to her brother about the books she had left earlier in storage. As of 1961, at age sixty-nine, she received $115 a month as Social Security retirement benefits, and was living in a small but respectable apartment on Hammond Street, a good neighborhood between West Hollywood and Beverly Hills. Her final residence was a small detached courtyard unit at 1032 North Hayworth Avenue, where she remained until her death in 1979. In 1965 she sent a Christmas card to Richter, in which she typed the following:

> With a wreath upon my door
> I have made one Christmas more;
> Garlands on my window greet

Everyone upon my street,
And the silent paper bell
Loudly rings dissension's knell.
Every candle, every tree
Lights a world of peace for me;
A cotton dove suspended flies,
And love looks out from tinsel eyes.

Not Lord Byron, perhaps, but also surely not the deranged ramblings of a woman whose brain had been ravaged by Alzheimer's for over a decade. Nor can one easily imagine these lilting words as the expression of a madwoman.

At the time of Margaret's institutionalization, treatments for mental illnesses—real and perceived—included some that now cause us to recoil in horror. With no effective drug therapies, patients were commonly subjected to electroshock treatments to reduce anxiety and neurosis, and not uncommonly given lobotomies. We do not know the details of Margaret's treatments; it appears that, if they were in fact of a barbaric ilk, they did not extinguish either her personality or her talent. It is possible that she was treated with one or more of the drugs that first became available in the 1950s, for example Thorazine, which had been introduced in 1954 and soon thereafter came into widespread use. Although many now liken the effects of Thorazine to those of a physical lobotomy, it appears that Margaret left the hospital with a greater degree of functionality than when she entered.

Margaret wrote enough poems during the early 1960s to fill a small volume that she apparently had printed herself. One is therefore left to wonder how much of Margaret Richter's earlier emotional turmoil might have been caused by a treatable metabolic disorder or imbalance. The medical establishment of the twentieth century was not generally kind to women whose chronic physical ailments could be perceived as a consequence of emotional problems, let alone those whose symptoms mimicked mental health disorders.

Following the trail to the next obvious question, one has to wonder to what extent Margaret's disorder(s)—rather than the tendency towards more fundamental mental instability—might have had a genetic component, and thus have contributed to Charles's problems as well. In his 1949 letter to his doctor, Richter commented on feeling ravenously hungry midmorning, unless he had eaten an especially large breakfast including eggs

Fig. 3.1. Small courtyard unit listed as Margaret Richter's last address. (Photo by the author.)

or meat. The doctor had suggested a certain type of diet that his wife had reportedly recognized. Richter does not elaborate on details, but one suspects a diet designed to prevent wild swings in blood sugar. To the modern eye, "breakfast including eggs or meat" certainly jumps off the page, suggesting that Richter's hunger pangs were mitigated by eating meals high in protein, and probably low in simple sugars. He reported that following the doctor's advice had decreased his hunger and general fatigue, "and my temper is better," he added. Perhaps, then, brother and sister shared the same tendency towards insulin resistance and diabetic, or at least pre-diabetic, conditions.

Another document in Richter's papers, this one a letter from Dr. Moriarty, reveals that his test results at the time were normal except for a cholesterol level of 100 milligrams per 100 cubic centimeters of blood. The doctor notes this to be a markedly low level, perhaps indicative of impaired liver function. Low cholesterol, a far less common problem than the other extreme, is now recognized as a serious medical condition in its own right. In 1999 Duke University psychologist Edward Suarez found that women with low cholesterol levels exhibited higher levels of anxiety and depression than counterparts with levels within the normal range. Cholesterol—a soft fatty particle in the blood—plays an important role in brain function, including the production of mood-stabilizing serotonin. At a time when high-fat diets and high-cholesterol diets represent a nearly epidemic public health problem, one can forget that cholesterol plays a vital role in human physiology. The substance is essentially a building block for cell membranes and many sex hormones.

Richter's health problems might or might not illuminate the nature of Margaret's problems. But whether the root of Margaret's problem lay in her blood sugar, her cholesterol level, organic degeneration within her brain, or something else, it is clear that Margaret Richter was enormously troubled throughout her life, and that she and her sister-in-law felt a violent mutual dislike for one another. Small wonder, then, that Richter and his young wife left the family homestead in Los Angeles on the occasion of his sister's return. The relationship between Richter's wife and sister may have been further complicated by the relationship between Richter himself and his sister. Of all of the enigmas in his enigmatic life story, perhaps none remains as mysterious as the relationship between this extraordinary man and his extraordinary sister.

To close out Margaret's own story, she remained in Los Angeles until she died on April 6, 1979. According to the doctor who attended her death, she suffered from cerebral arteriosclerosis for the last five years of her life, a condition in which the arteries in the brain thicken and harden, frequently leading to dementia. A final cerebral thrombosis—a stroke caused by a blood clot—on April 5, 1979, was the ultimate cause of death. She also suffered from rheumatoid arthritis. Her caretaker in her final years, a woman by the name of Hildegarde Pigorsch, died in 2005.

Virginia Woolf imagined for the sake of illustration that the most famous writer of all time, William Shakespeare, had a sister. The most famous seismologist of all time did have a sister: her name was Margaret Rose. She did not share her brother's analytical talents, but her talent for verbal expression far surpassed his. It stands as testimony to these talents that she was able to establish herself as a poet and a university teacher in the early decades of the twentieth century. Yet one is left to wonder about the battles that she faced: the possibility that chronic—perhaps even treatable—physiological problems were chalked up as only so much female hysteria or neurosis. History records far less of Margaret's life story than that of her more famous brother's. We can be thankful that Margaret not only escaped Judith Shakespeare's fate but also, unlike the George Eliots (pseudonym of

Mary Ann Evans) of earlier times, established a writing legacy in her own name. Still, one would love to read more.

Margaret's death merited two brief lines in the "Vital Records" section of the *Los Angeles Times.* By the end of her days Margaret's children of the mind proved to be poor company indeed; her lonely days had, as she had foreseen decades earlier, grown lonelier still. Her only living relation, her younger brother, had looked after her—had loved her—until the end, but by 1979 struggled to take care of himself. Just another addled old woman; just another nearly anonymous death. No memorials to a life of accomplishment; no recognition of a star that had once shined brightly in its own right. Virginia Woolf suggested that "Anon." was a woman; that, in fact, women lack the force of ego to carve their name on every signpost. Yet Margaret Richter did care about signposts: for decades she carefully cut and pasted into a scrapbook not only her published poems but also newspaper clippings about her work. She had a lifetime of half-hours to call her own; a lifetime of rooms of her own. Perhaps her failings, then, were only her own. Perhaps not. One wishes one could read more.

CHAPTER 4

Harnessing the Horses

EARTHQUAKE

I set my aspiration on the soundest rock,
And chose my building-stone with care:
No moral clay, no pious wooden block,
But granite fact, and rigid logic layer on layer.

I built high towers, not of ivory, but stone;
Wide rooms for books and serious things.
There was house-room for solid work alone,
But on the top arose my best imaginings.

Finished at last, I felt an architectural pride;
No mind had such a house before.
Then, as I was about to march inside,
There came a violent shaking and a stunning roar.

In spite of argument the clashing stones broke free;
In vain the rock stood firm and sound.
My towers collapsed in streams of masonry,
And all my lofty dreams fell crashing to the ground.

Though not beyond repair, the splendid house was
wrecked.
I cursed, and left it unrestored.
Puzzled, not comprehending my defect,
I came with my perplexities before the Lord,

Who, smiling, said, "You are no mason, it appears.
Return, and make a new assault.
Use better mortar, and dismiss your fears.
The rock and stone were good; the builder was at fault."

—*Charles Frances Richter, July 2, 1933*

AT THE AGE of twenty, armed with a Stanford degree and at the point at which a young man prepares to take on the world, Charles Richter fell apart. He had begun graduate studies at Stanford but lasted only a year before his anxieties got the best of him. From his first venture into formal schooling—his first real steps into the outside world—Richter had never functioned well in society; on the brink of adulthood he found himself unable to function at all. He was not—and probably never would be—the man he yearned to be, that much he knew; far less certain was the answer to the question, what kind of man was he?

Richter wrote virtually nothing about his year in a sanitarium, during which time he was under the apparently capable care of Dr. Ross Moore. His initial forays back into the outside world—respectable but not overly demanding jobs—must have helped him gain traction on the road to the rest of his life. With money saved from these jobs he was able to finance his return to graduate school in 1923.

The roots of the California Institute of Technology, where Richter returned to graduate school, date back to 1921. The university was not created from whole cloth, but rather grew from the Throop Polytechnic Institute, founded in 1891 by wealthy former abolitionist and Chicago politician Amos Throop. In the early years of the twentieth century Throop enticed a trio of eminent scientists—astronomer George Ellery Hale, leading physical chemist Arthur A. Noyes, and physicist Robert A. Millikan—to spend a few months a year visiting his institute in Pasadena. World War I intervened, and all three scientists went to Washington, D.C., to organize and lead scientific studies for the war effort. By the time the war ended, the formidable trio was determined to put American science, and the Throop Institute, on the map. By 1921 the school had a substantial endowment, a new educational philosophy, and a new name.

Hale and Noyes persuaded Millikan to leave the University of Chicago and take the position of director of the Norman Bridge Laboratory of Physics; he also became the administrative head of the institute as a whole. By this time Millikan had attained enormous stature as a physicist. His famous "oil-drop experiment," which for the first time measured the charge of an individual electron, had been done in 1910–a colossal achievement for its day. In 1923 this experiment, which a half-century later could be repeated easily by teenagers in a rudimentary high school physics lab, earned Millikan the Nobel Prize. Taking the reins at Caltech, Millikan and his colleagues soon lifted Caltech into the top echelon among American research centers. Millikan also became a public figure, an active spokesman for science and science education.

News of Millikan's arrival reached Richter, a stone's throw away from Pasadena in downtown Los Angeles. Richter later explained that he "couldn't miss the opportunity to hear his lectures. The result was that very soon I gave up my employment and started Caltech as a graduate student."

Richter describes Millikan's lectures in somewhat surprising terms: "very well organized," but also "occasionally somewhat slow." On occasion Millikan's presentations were followed by lectures by Professor Paul Epstein, who focused on recent developments in theoretical physics. Richter initially found Epstein difficult to follow, not only because of the physicist's degree of mathematical rigor but also because of Epstein's heavy German accent. After enrolling at Caltech as a graduate student Richter found himself in Epstein's classes, by which time both the teacher's English and the student's mathematical prowess had improved. By his own account, Richter "began to get a very great deal out of his instruction." Among Richter's papers at the Caltech archives one finds rows of notebooks in which the young student transcribed lectures, including Epstein's classes, verbatim. Richter's notes are astoundingly complete, including even his professors' opening and closing remarks. One marvels that he managed to write that fast. Or perhaps he knew shorthand, and transcribed the lectures in full later.

Richter found himself at Caltech during heady days. "This was just about the time," he later said to interviewer Ann Scheid, "when quantum mechanics was evolving and the whole atmosphere of atomic mechanics was changing." Guest lecturers arrived routinely on the Pasadena campus, the likes of Erwin Schrödinger and Hendrik Lorentz. Neither the guest lecturers, however, nor Millikan himself captured young Richter's attentions the way that Paul Epstein did. "He was a very beautiful lecturer," Richter explained, "in that his lectures were always carefully planned and organized. He had a number of odd mannerisms, some of which were Germanic and some of which were individual. I remember he was something of a pacer, and there was one particular lecture room which had a loose board or something at one end of the lecture platform, and he almost invariably hit that with a plunk. I'm not sure whether it was completely an accident."

Richter's thesis topic was suggested by Millikan rather than Epstein. Millikan gave him a paper describing the hypothesis of a spinning electron, which seemed to reconcile a number of contradictory previous results. "Would you look it over?" Millikan asked. Richter "found that indeed it promised to be at least a partial theoretical answer to some of the matters that were troubling [Millikan]." From this encounter Richter's thesis topic was born: further work on a hydrogen atom with spinning electrons. The thesis was carried out under the supervision of Epstein rather than Millikan, for reasons that are not altogether clear. It is possible that Millikan's responsibilities left him little time to supervise graduate students, but clearly Richter also felt a sense of professional—and maybe to some extent personal—affinity with Epstein.

By Richter's account, however, he never came to know Epstein well outside of their professional relationship—although he did on occasion refer to his former mentor as "Eppy," which suggests a certain kinship. Richter did, however, establish a close friendship with a student who once distinguished himself by pointing out an error in an equation during one of Epstein's lectures: Boris Podolsky. Podolsky went on to a measure of fame, at least in physics circles, for having coauthored a paper with his postdoctoral advisor, Albert Einstein, and fellow postdoc Nathan Rosen, at the Institute for Advanced Study at Princeton. The article, "Can Quantum Mechanics Description of Physical Reality be Considered Complete," posed what is today known as the Einstein-Podolsky-Rosen (or, simply, EPR) argument, which soon became the centerpiece of the debate about the implications of quantum theories. The paper answered the titular question in the negative, that quantum mechanics was not in fact complete in the sense of being able to reconcile inherent paradoxes.

Podolsky played an important role in the development of modern physics; he also inarguably played an important role in Charles Richter's life. It was through Boris that Richter met his future wife, Lillian Brand. The pair met at Podolsky's home in 1927, where Lillian was living with Boris and his wife, and "helping them out," in Richter's words. By this time Lillian had separated from her first husband, Reginald Saunders, although their divorce was not yet final. A domestic "au pair" position might well have been the best Lillian could do for herself. Richter did not say much about their early encounter beyond the fact that Lillian "considered herself very exceptional, being the native daughter of a native Californian. There aren't many of those." By Richter's later account, Lillian's grandfather had built the very first house in the city of Pomona, east of Los Angeles.

Richter's friendship with Lillian blossomed. On July 19, 1928, two years after his grandfather's death and after Lillian's divorce became final, she and Charles were married. Lillian had a son from her previous marriage, Reginald "Butch" Floyer Saunders Jr., born in 1925. The boy did not live with Charles and Lillian; in the 1930 census he appears as one of two young "boarders" living in the home of Helen Meany, at 1533 Orange Avenue in Los Angeles. One can only assume this to have been a foster care arrangement. The 1930 census lists Saunders's parents as unknown, but Lillian clearly maintained contact with her son. Richter later described his relationship with his stepson as close: during the latter half of Richter's life the two men were occasional hiking companions. According to Bruce Walport, the son of Lillian's younger sister Ethel, Butch did not remain within the

Fig. 4.1. Lillian Brand Richter, circa 1925. (Photo courtesy of Laurie Walport.)

foster care system but grew up with his father, Reginald Floyer Saunders, and his stepmother in the San Francisco Bay Area.

Reading the account of Ricther's family life, one might pause to reflect on the changing face of American families through the twentieth century. While divorce had become commonplace by the closing decades of the twentieth century, it was far from common in its early decades. Even in 1960 the number of divorces per 1,000 American married women in a given year was a mere 9; in 1900 the number was approximately 4. Put another way, in 1900, only 1 out of every 250 married women in America divorced in a given year. Yet Richter's mother had been a statistical oddity—twice over by her own account—before the nineteenth century had come to a close, and Richter's wife was one as well. Richter's sister, meanwhile was a part of a different small minority during the first half of the twentieth century: women who never married. When Charles Richter spoke of "his sort of women," clearly he referred to a breed rather different from the norm.

In later interviews Richter spoke little of personal matters, including his marriage, beyond the basic facts: who, what, where, when. His soul-baring 1949 letter to Dr. Moriarty, however, tells us considerably more. The letter tells us that two of the three women in Richter's life—his sister and wife—could not stand each other.

The years after Richter's emotional breakdown at age twenty and before his introduction to seismology at twenty-eight suggest a life that, although derailed for a time, was more or less back on track by the time he reached graduate school. Through the 1920s Richter enrolled in a Ph.D. program and completed his degree, hobnobbed with some of the most brilliant scientific minds of the day, and met and married an apparently suitable young woman (albeit one with a somewhat checkered past). Richter had had more than his share of coming-of-age anxieties, but clearly, or so it seems, they were safely behind him by his Caltech years. His *Los Angeles Times* obituary certainly suggested as much: "'He had a brilliant mind,' said one former colleague, 'but it bordered on instability in his younger years.' Richter sought psychiatric help in the 1920s, something of a daring thing to do then." (The obituary goes on to say that it was his psychiatrist who introduced Richter to his future wife, which, as we have already seen, is contradicted by Richter's own account. Richter does provide a clue to the misunderstanding: for a while he found himself "talking with Podolsky about myself, and we have even undertaken some use of psychoanalytic methods—with no very great success, although some points have been cleared up.") Nevertheless, the impression from the obituary is clear: the instabilities plagued Richter only, or at least chiefly, during his younger years.

This is, perhaps, the portrait of Richter that the outside world saw by the end of the 1920s. It is, without question, a vastly different portrait than the one Richter himself drew through his graduate school years and beyond. One need not wonder what was going on in his head, so to speak, during these years: he wrote much of it down in a personal diary dated June 1926 to June 23, 1928. The scrawled handwriting of the journal does much to illuminate Richter's state of mind during the concluding years of his Ph.D. research—a state of mind clearly still wracked by turmoil and instability.

Most of the journal appears to be a simple diary, written to nobody but himself. On early pages of the journal, dated June 20, 1926, Richter paints a portrait of discontent: "It must certainly appear, to anyone less fortunate than myself, that I have ample occasion for contentment," he wrote. "I am successfully engaged in a research which, although not profoundly fundamental, is at least in the front of present scientific advance—certainly as much as any young and relatively inexperienced physicist has any justification in expecting. I am in the good graces of two of the leading personalities in my subject, and am looked upon as a promising young worker."

And yet, "with every good thing thus steadily coming my way," he went on to say, "I cannot settle down and enjoy the advantages of this admirable position of mine." He then proceeds to outline the dilemma that occupied his thoughts—the dilemma that would plague him for the rest of his days, which even his closest scientific colleagues never glimpsed. "All my training," he wrote, "and all my natural aptitudes run in this direction, and still I find myself continually struggling away from it. I cannot make up my mind wholly to give myself up to scientific investigation; for I am always brought up short by an indefensible hankering after artistic expression. For me this means literary expression, since the other arts all demand a more or less manual skill which I find is completely alien to my natural personality."

There you have it. The most famous seismologist of all time was not only a physicist rather than a seismologist at heart; he also yearned deeply to express himself as a writer. Looking back, Richter attributes his earlier emotional breakdown to a "surfeit of scientific occupation." "It was," he wrote, "the realization of the artistic, or as I called it then, the spiritual aspect of the world, which first raised me out of that depression, and it was the final accomplishment of self-expression in poetry which at last permitted me to return to my work." Whatever other therapy techniques Dr. Ross Moore employed, he encouraged Richter to deal with his demons by putting them down on paper. Having established such an outlet, Richter ironically found himself able to return to his previous scientific occupation, which he then found to be at odds with the very artistic expression that allowed him to work in the first place. "Now," he wrote, "after continuous scientific work, I am quite naturally in a cooler and nervously exhausted state; and I find my faculty of artistic expression at least blunted, if not completely lost."

This last turn of events might have alarmed a young man who had come to see poetry as salvation, but by this time Richter also realized that "one of the principal elements in my former collapse was a failure to comprehend that branch of science in which my natural abilities correctly belong. I know myself now for a theoretical physicist; but it has taken five years of uncertainty, and the judicious advice of Doctor Millikan, to finally place me in that position."

Thus the journal begins to hint at the extent of Richter's struggles with not only self-expression but also self-awareness. He had some inkling of the nature of what tormented him and clearly yearned to understand it more fully. Yet reading words penned in penmanship that struggles valiantly to maintain a decent minimum of orderliness, one quickly gets the impression that it was no easy task for Richter to take measure of the man that he was.

Notwithstanding his self-described congenial home in the field of physics, a key element of Richter's continuing emotional turmoil clearly lay with his desire to reconcile what he saw as disparate parts of himself—the part that felt drawn towards science with the part that continued to yearn for artistic expression. Certainly he despaired of ever uniting the two: "It seems to me as though the hopes I entertained then, of uniting poetry and science in my person, have very little chance of being realized. Such an achievement demands a great genius, and that genius I do not seem to possess."

"If it is so," Richter continued, "it is good to have measured myself, to recognize my limitations; but I am not certain, and assuredly cannot be certain, that the conclusion is right. It does not necessarily require a towering genius to make such assimilation as I had in mind; it should apparently only require susceptibility both to science and art and that, but a fortunate chance of environment, I have." He immodestly adds, "And even if genius is required, I may have it. I am still young and do not know my limits. I know I am not physically strong, nor practically minded, but I cannot convince myself that I lack artistic possibilities. In fact, I have good witness to the contrary, in several promising productions a short time ago."

If, as seems certain, Richter had no inkling that his emotional struggles likely had neurobiological roots, and no understanding that they interfered with the very expression that he yearned so desperately to undertake, he surely had no sense that his demons were also furiously at work at the very core of his being, working against the unification of which he spoke. Another man might have been happily content to pursue a career of scientific research during the week and write poetry—or take photographs, or paint portraits—on the weekends. Richter was, instead, given to an intensity of focus that drove him relentlessly—drove him to devote his whole self to one commitment or the other and, if such was not possible, to struggle to reconcile what he saw as deep philosophical inconsistencies between very different callings.

And struggle he did. Much of the rest of the journal is filled with Richter's deeply philosophical musings. "The conflict in my mind between the rival claims of science and art is of course no merely personal affair." Thus begin long ruminations on the nature of science, the nature of art; the conflict between life and matter; the nature of the self . . .

> Of art in all ages we have a reasonably consistent idea, however far we may be from precise definition. But in determining the relations of science to art in past ages we have to be very careful to what was the science of that age, on which the art was based . . .
>
> As a matter of fact, in a thoroughly modern mind, the territory of religion is almost entirely preempted by science, history, and applied science. Accounts of the origins of things are historical science; explanations of phenomena are science proper . . .
>
> My message to anyone seeking anchor in the present chaos is this: the world is the self. Your desire for unity is the assertion of the unity of your personality. Make that the central fact of your mental life, and cling to it.

(It is possible that Richter would have appreciated the ancient practice of yoga.)

> It seems I have now got at the bottom of my sentimentality—the effect which old songs . . . have always had on me. I got at it by noticing that the effect on me of Keats' 'Forever shalt thou love and she be fair' is of the same species. The fact is that all these things are associated with love-making—with physical love-making, in fact. Now I have had none of that, and all those impulses in me are frustrate; the result is that I feel a certain disappointment when such associations are brought up, and I dissolve in tears.

It is possible that among Richter's more unusual tribulations there were some rather ordinary ones as well. This passage suggests that Richter was, at age twenty-seven, still a virgin. He did write elsewhere of a first love affair earlier, at age twenty-one; however it is not clear how far this early affair progressed.

Richter's reflections are exhausting to read; one can only imagine what it must have been like to inhabit the mind that wrote then. He went on to reflect in detail on the nature of human personalities: introverts versus extroverts, male versus female temperaments, male versus female introverts, male versus female extroverts. He further factored in the degree of sexuality to construct an elaborate table: female, strongly sexed extrovert; female, strongly sexed introvert; female, less sexed extrovert; female, strongly sexed introvert. And so forth.

He considered at length the changing role of women in society. One unusual journal entry begins with the heading, "Letter mailed Dec. 20"—no other entry begins with anything other than a date. Richter does not identify the recipient of the letter, but what he writes makes clear the fact that his reflections were written to be sent:

> I was quite prepared to wait for your answer; I felt that it would be very kind of you to write at all. In my turn I have had to reflect for a while before I could continue intelligently.
>
> If I may, I should like to go on with the discussion; for I have been very much interested, both by the statements you made and by those which you did not.
>
> . . .

He continues,

> As I understand you, you believe that all this concern about such an ethereal matter as the relation of science to art is a little superfluous when our society is confronted with so many vital problems, so many flagrant injustices; that is why you are a communist, not a scientist or an artist. Am I wrong?

Richter goes on to opine on the subject of the ideal society, referring to the "Russian experiment" and noting that he "has no desire to see that experiment repeated here, though its results (when they are reached) may well be applied everywhere." He offers his own vision—of sorts—of an ideal society: "My hope for the transformation of society lies chiefly in the power of art."

Eventually he addresses the issue of a woman's role in the world, a subject that he considers at length. Many years later an interviewer would ask him about the small number of women in sciences: had any women ever been involved with the Seismo Lab in a research capacity? Can you tell me about the controversy regarding the initial admission of women at Caltech?

Richter's answer to that last question proves interesting: amid an apparently heated discussion on whether or not the institution should admit women, he observes, "the thing that finally made a crack was the arrival of a new staff member in chemistry who insisted on bringing his assistant with him or else he wasn't going to come . . . a woman graduate assistant who had been practically indispensable to him in his research." Yet Richter's answers to these questions runs strongly towards the perfunctory: one does not come away with the impression that he ever gave the matter a great deal of thought.

Consider, however, his words from a half-century earlier:

> Even supposing that women have hitherto contributed nothing to the mind of the race, I should not think that any guarantee of the future. Women have been ostensibly worshipped as superior beings, and actually oppressed as inferior ones. Those having real intellect come to maturity in a world where all the thought-forms are masculine invention. Is it any wonder they have produced so little? Whatever doubts you may have personally, you know quite well that any man who has any delicacy of feeling recognizes, in a fine woman, a keenness of intuitive perception far in excess of his own. The failure of that perception to express itself is the fault of the world, not of the woman—or so I think.

His reflections continued,

> Please don't imagine that this is simply an indiscriminate adolescent idolization of women in general. I have known "respectable" women who were about as much use in the world as a rat-flea; and others, not so "respectable," who could accomplish more good in a single evening than most men manage to bring to pass in all their days. I honestly do consider a really intelligent woman as a more important personage than any man (emphatically more so than myself), simply because in the present state of affairs she has more opportunity to do something permanently worth while.

An interesting mix, certainly, of condescension ("Is it any wonder they have produced so little?"), enlightenment ("come to maturity in a world where all the thought-forms are masculine"), and, perhaps above all, reflective thought. It remains an interesting question, to whom was this letter written? A number of passages point to a woman, but who? Not his future wife: they had not yet met. The opening paragraph of the letter suggests that it was not written to his mother or sister—the only two women known to have been a part of his life at the time: "A correspondence like this," Richter wrote, "is necessarily slow if it is to mean anything. I was quite prepared to wait for your answer; I felt it would be very kind of you to write at all." Presumably one does not begin a letter to a close relative thusly. There is, moreover, no evidence that any of the women in Richter's lives were communists, as the recipient of the letter clearly was. Throughout his life Richter apparently initiated any number of correspondences with people with whom he was only vaguely acquainted, or even not acquainted at all. It appears that his congenial female communist correspondent was one of them.

Whoever the recipient was, the letter provides one more measure of illumination into Richter's inner thoughts as he worked towards the completion of his Ph.D. thesis. By his own account he remained in good standing with two leading scientists in the rapidly developing field of theoretical physics; by his own account his work was progressing well, and, if not worthy of the Nobel Prize, was at least successful research at the forefront of his chosen field. And yet, he wrote, "I cannot settle down." The pages that follow make this observation sound like an understatement of biblical proportions.

By his own account, Richter had mostly stopped writing verse by the closing years of his grad school tenure, and felt emotionally drained—"in a cooler and nervously exhausted state of mind"—by the demands of an intense scientific research career. As one can easily enough imagine, his marriage also provided an outlet for some of his previously unharnessed—and potentially self-destructive—energies. In having found graduate school to be a consuming, emotionally draining endeavor, Richter is far from unique. In his case, however, a wellspring of nervous energy and emotional turmoil perhaps needed to be drained.

Even in an emotionally spent state Richter did have enough energy for some poetry during these years. Richter's continuing struggle shines through the poem "Outcry," dated July 2, 1926:

> In page on page of symbols I rehearse
> The mathematics of the universe;
> On one sheet set effect, on one sheet cause;
> Here draw the atom, and there write its laws.
> Insulted nature takes a vengeful toll,
> For mathematics penetrates my soul.
> No more extravagances can I find;
> Order, and only order, rules my mind.
> I see the world immutable at core;
> I find my changing fantasies no more.
> Must I perform unwilling sacrifice,

And offer up my heart clamped in a vise?
Powers eternal, stoop to pity me:
Let me both love and know; let me go free!

In the opening stanza of another poem, "Synthesis," dated November, 1926, Richter wrote of his continued need for expression against the confining forces of scientific thought and rigor:

It is too much, and I must write.
The mind that stares at truth all day
Finds that a harsh, unyielding light
Steals life and hues and breath away.

He wrote other poems during these years, including several about the mountains that, through these years as well as other times of turbulence, provided his one true sanctuary.

His journal reflections also make clear the fact that, even in a nervously exhausted state of mind, Richter was thinking about more in a day than some people reflect on in a lifetime. His formidable intellect grappled with such questions as the nature of art and science, the nature of the self, the essence of human personality, the essence of society.

His words during this turbulent chapter of his life suggest that he had some inkling of the tribulations that he struggled with. Certainly he yearned fervently to understand them more completely. "I think," he wrote on April 20, 1928, "that I have now reached a point at which I can be reasonably certain that my chief artistic ability is definitively one of expression and that the problem is not simply that of having nothing to express." The biographer is tempted to put an exclamation point at the end of this sentence.

> I feel that I have four essential ideas, or groups of ideas, which differ at one point or another from those of the majority of intelligent persons, and which taken together constitute my intellectual individuality. It is the proper expression of this individuality in verse or perhaps prose that makes my present problem. I wish to express, to communicate—
>
> **I.** The need for a system, a synthesis, a philosophical outlook.
>
> **II.** The purpose in living—the need to contribute—by creation or procreation
>
> **III.** The necessity and significance of art (literature and music)
>
> **IV.** The need for a proper appreciation of the viewpoint of modern science.

Having taken measure of himself thus on April 20, he writes his next journal entry as, simply, "June 8, 1928. Charles Frances Richter, Ph.D." And here the journal ends.

Having attained his Ph.D. in the years before the depression, Richter was unmarried, by his own account, "pretty well free," and not in any particular hurry to move on. He further admits to not having given much thought to what he might do after attaining his degree in physics. He later told an interviewer, "What was in the back of my mind—of course, I had already done some work for the Institute as a student assistant, which hadn't turned out too well. I had the feeling that I was in the good graces of the administration, enough so that if I stuck around, probably something would be found, and I might eventually work into a permanent position, because I had demonstrated interest and ability in this very critical field of quantum mechanics."

In the end, the mountain did come to Mohammed, but not the mountain Mohammed had been waiting for. In 1927 Richter was summoned by Robert Millikan: the nascent Seismological Laboratory was looking to hire a part-time research assistant with a background in physics. Was he interested?

The rest, as they say, is history.

By Richter's later account, he was content to take a position outside of his chosen field "because so long as I could stay in or near Pasadena I could keep in touch, which I did to a certain extent." Certainly he had no intention of leaving physics altogether: "I had indeed some problems in theoretical physics in mind which I wanted to work on and see whether I could get somewhere with them, which proved not to be the case." Richter did, in fact, send one highly mathematical calculation concerning the energy levels of helium atoms to Caltech physicist Harry Bateman in 1929, to which Bateman responded with interest and encouragement. This interaction apparently did not, however, blossom into the kind of collaboration that Richter had hoped for. At the same time, he was clearly happy to have found employment that allowed him to stay in Pasadena: "I didn't feel I was departing," he wrote. "I did want to keep in touch, and the work at the laboratory I felt at least provided me with the means to stay around here instead of taking a position in some other part of the country." Reading these words, one is inclined to suspect that "some other part of the country" would have been unattractive to Richter for reasons beyond those stated. Connected to his birthplace by only the whispery memories of young childhood, Richter's unhappy experience at Stanford can only have cemented whatever innate homebody tendencies he possessed. The Los

Angeles area was home, the center of the universe for a nuclear family with especially tightly bound orbital shells—family members did on occasion leave, but never for long, and never to successful ends.

In fact, as we will see later, Richter's limited turning radius throughout his life was likely a consequence of his nonstandard brain wiring: biology had left him ill-equipped to deal with the kind of change that other people take in stride. And what was difficult for most people could be nearly impossible for Charles Richter. Here again one finds an unfortunate trait for a budding young scientist to possess. Moving cross-country might not have been as common in the early decades of the twentieth century as it became later, but academics have always been, by necessity if not by inclination, a breed of gypsies. In the 1920s as now, a young scientist in Richter's position does not find a job by picking up the hometown newspaper and looking for classified ads that say, "Seismologist wanted."

Richter, however, found himself reluctant to make another stab at establishing a life outside of Southern California. One wonders if he didn't find himself simply lacking the wherewithal to make life-changing decisions. Beating the bushes to find academic employment is no easy task for any young scientist. It requires a healthy measure of intestinal fortitude, assertiveness, confidence, and the ability to develop social and professional contacts. In all of these regards, Richter's complex temperament would have worked against him. First and most obviously, a man like this would have found "beating the bushes" far more difficult than the typical, confident, perhaps somewhat brash young scientist. Second, the fact that he did not fit the standard bright-young-scientist mold would have almost invariably colored the scientific community's perceptions of his talent and potential.

Richter was, however, a man of substantial talents, and found himself at Caltech at a time when the seismology program was beginning to get itself off the ground (so to speak). The academic community might not have welcomed this young scientist with open arms, but it did find a place for him. Nearly buried alive by the collapse of his own carefully constructed towers, he had indeed returned to make a new assault. Thus did Charles Francis Richter, for the first time in his life, turn his attentions to seismology.

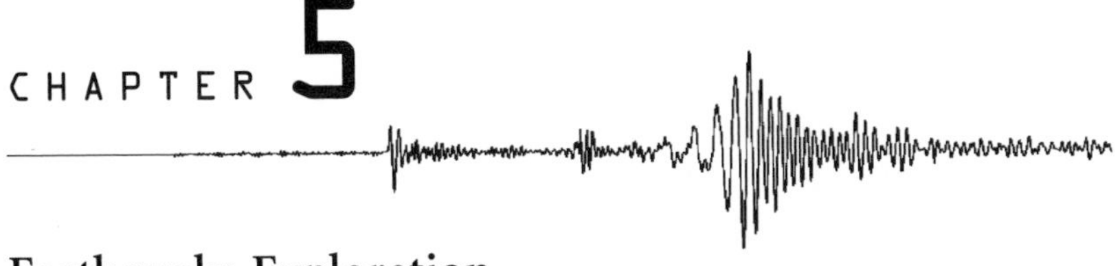

Earthquake Exploration

Seismology owes a largely unacknowledged debt to the persistent efforts of Harry O. Wood for bringing about the seismological problem in Southern California.
—*Charles Richter, National Earthquake Information Service interview*

LIKE any life story, Charles Richter's must be told in the context of the larger drama in which he played a leading role—in Richter's case, the business of earthquake exploration in Southern California. When he first stepped into this drama, the play was still in its opening act, but the scene had already been set. This chapter steps back from Richter's life to tell that part of the story: the earliest days of earthquake exploration in Southern California.

Earthquakes were a familiar part of life in the United States as a whole even before the country was a country. On November 9, 1727, an earthquake struck near Newbury, Massachusetts, shaking down nearby stone walls and chimneys. The largest historic earthquake in New England, on November 19, 1755, struck east of Cape Ann, and was felt as far as Nova Scotia to the north and Chesapeake Bay to the south. Over the winter of 1811–12 a remarkable sequence of earthquakes, with three distinct mainshocks and thousands of aftershocks (some quite large), struck the midcontinent near the present-day bootheel region of Missouri. If Americans today view earthquakes as a California problem, the so-called New Madrid sequence of 1811–12 left nineteenth-century Americans with, ironically, a better appreciation for earthquake hazard as a national—indeed, a global—problem. This sequence included four temblors now thought to have been upwards of magnitude 7, and many thousands of aftershocks. A large and destructive earthquake near Charleston, South Carolina, in 1886–

now thought to have been of magnitude 7–further cemented Americans' appreciation of earthquake hazard.

Yet by the time the nineteenth century drew to a close, most Americans—those of a geological persuasion in particular—had a growing awareness that earthquakes in the eastern and central United States are not nearly so common as those in the West, in California in particular. Geologists began to realize that California had had more than its share of large temblors: major earthquakes in 1812, 1857, and 1872 had caused only modest damage and loss of life only by virtue of sparse population density. A large Bay Area earthquake in 1868 provided a rather more substantial societal jolt, striking in the heart of a burgeoning economic hub.

By 1900 the awareness of California earthquakes had grown to the point that some in the region began to actively downplay the idea that California was earthquake country, in particular those with vested interest in the continuing development of the Golden State. No sooner did scientists begin to understand the severity of California's earthquake problem than business interests began to feel threatened by it. Who would want to live in, much less invest in, a state that couldn't sit still? Without question those interests took notice of the earthquake that struck on October 21, 1868. Scientists now estimate that this temblor had a magnitude of 7 and occurred on the Hayward Fault. Most of the buildings in the city of Hayward suffered extensive damage; in nearby San Leandro the courthouse and city jail collapsed. Less severe damage was reported in towns as distant as Santa Cruz to the south and Santa Rosa to the north.

By 1870 the gold rush was two decades old and total population of the Bay Area, including relatively far-flung counties such as Sonoma and Santa Clara, had grown past the quarter-million mark, with nearly 150,000 in San Francisco alone. The California boom had begun: in the decade following 1870 the Bay Area population would nearly double. Explosive growth meant explosive business opportunities, and business opportunities meant business interests. Such interests did not take kindly to threats, and chief among these threats were pesky natural hazards such as the 1868 temblor. By that time the temblor could not be dismissed as a fluke: other moderate shocks had rocked the greater Bay Area in the closing decades of the nineteenth century. Scientists now look back at this spate of moderate shocks as a possible harbinger of the great earthquake that rocked the state in 1906.

In a letter written to the Seismological Society of America several decades after the 1868 earthquake, seismologist George Davidson discussed pressures that arose in the aftermath of the 1868 quake. This letter was not published until 1982, when geophysicist William Prescott found it and discussed it in a note in the *Bulletin of the Seismological Society of America.* In his letter, Davidson tells of the five subcommittees that had been formed following the 1868 temblor, three to investigate the performance of structures, one to summarize scientific results, and one devoted to legal matters. The news media eagerly awaited the report; news articles in January 1869 expressed criticism of the slow progress of the effort.

Among the findings of Davidson and his colleagues were some that are very familiar to modern ears: most of the estimated $1.5 million property damage occurred on "made land" and other sites where buildings had been built atop loose sediments and soils. (By "made land," Davidson referred to areas where artificial fill had been used to develop waterfront land that was not otherwise buildable.) In Davidson's words, "Report was carefully prepared but [committee chairman] Gordon declared it would ruin the commercial prospects of San Francisco to admit the large amount of damage and the cost thereof, and *declared he would* never publish it." In Davidson's estimation, "the earthquake of 1868 was more violent and destructive than that of 1906," the degree of damage due to the "suddenness and extent of the movement." Regarding the missing report, Davidson added, with understated but obvious regret, "I know of no one who had taken a copy of the report" prior to George Gordon's apparent decision to suppress it. Only a handful of the report's conclusions ever saw the light of day: we know about them today only by way of Davidson's letter.

As late as the opening years of the twentieth century, many in California sought, with at least some measure of success, to paint a big happy face over the Golden State—its substantial earthquake hazard in particular.

On the morning of April 18, 1906, the earth itself put an end to most (although not quite all) such nonsense. The great San Francisco earthquake, as we know it today, is now estimated to have been of magnitude 7.8 to 7.9–a whisker shy of the great magnitude 8 earthquake that today represents the proverbial Big One in the minds of most Californians. The temblor left a jagged scar on the San Andreas Fault from San Juan Bautista to Point Reyes, a distance of 430 kilometers (270 miles). Shaking was strong enough to break strong, mature trees in two along a remote swath of the fault through the Santa Cruz Mountains, and caused well-built masonry buildings, for example on the Stanford campus, to crumble. In San Francisco the shaking caused substantial damage, once again especially on the "made land" where damage had been concentrated in 1868. (Some lessons are destined to be learned and relearned. Or rather, learned and forgotten.) Far worse than the damage from shaking, however, were the fires ignited by ruptured gas mains and other sources—fires that a badly disrupted water supply could not hope to extinguish. The city burned, very nearly to the ground, over a staggering extent of its acreage.

And still, even in the aftermath of horrific destruction, business leaders sprang into action to downplay the earthquake risk. The disaster was portrayed as a fire, and in a sense this was true; the fire had done far more damage than the actual shaking during the earthquake. The fact that if not for the earthquake, there would have been no fire—that was incidental. When seismologists proposed adding earthquake provisions to the building code, they were pitched to officials not as "earthquake provisions" but as "wind provisions."

Following the 1906 earthquake scientists mapped not only the full extent of the break, or rupture, but also most of the San Andreas Fault. The 1906 break extended from San Juan Bautista in central California to near Point Arena. Somewhat remarkably, in the weeks and months after the earthquake, geologists followed the trace of the fault over hill and over dale, through remote and sometimes rugged stretches of central and Southern California, as far south as San Bernardino. In this undertaking scientists were guided by a trail of geologic breadcrumbs: the scar left by a great southern San Andreas earthquake in 1857, which by 1906 had not been entirely erased from the landscape.

By the very early twentieth century geologists were thus aware of the southern San Andreas Fault, which runs along the northern edge of the San Gabriel Mountains to the north of the greater Los Angeles region. Yet still the question remained: to what extent were earthquakes truly a problem for the growing population centers of Southern California? The southern San Andreas Fault does not, one must note, bisect the greater Los Angeles metropolitan region, but is a good forty kilometers away from even its northernmost valley communities. For earthquake hazard purposes, distance is a godsend: shaking is strongest in proximity to an earthquake and diminishes rapidly as distance from a fault increases. Perhaps, some concluded, the major metropolitan regions of Southern California did not share the same degree of earthquake hazard as the Bay Area to the north, which was known to be bisected by the clearly active San Andreas and Hayward faults.

Charles Richter would later recollect the 1906 disaster among his few, fragmentary, early childhood memories. He was a very young boy at the time, far removed in every sense from the drama that played out in California in the aftermath of the temblor. Many decades later, however, in his 1979 interview with Ann Scheid, he clearly knew the story, chapter and verse.

By Richter's account, the 1906 temblor convinced the U.S. government that earthquakes were a serious problem. At the time, the only government agency that dealt with earthquakes was the Weather Bureau, and then only in a most informal sense: weather observers were instructed to report anything unusual, including earthquakes. In its early years, the mission of the U.S. Geological Survey focused on mapping mineral resources, not researching and mitigating geologic hazards. U.S. Geological Survey scientists had, however, taken the lead on a couple of notable earthquake reports. A hundred years after the 1811–12 New Madrid earthquakes, USGS geologist Myron Fuller published the first comprehensive scientific report on the earthquake sequence. Following the 1886 Charleston earthquake, geologist Clarence Dutton authored a remarkably thorough report that was published by the USGS. Although devoid of modern seismograms—the instruments barely having been in existence in 1886–the report documented the effects of the temblor in impressive detail. (To be precise, Dutton was not a U.S. Geological Survey scientist per se, but rather an army officer who managed to get himself detailed to the geological organization during its early years. He later managed to get himself exiled to a munitions depot in San Antonio, but therein lies another life story altogether.)

Notwithstanding these early efforts, "the Survey," as it is now known in scientific circles, did not run any earthquake monitoring programs in its early years. Nor was any other government agency involved with substantive earthquake monitoring efforts at the start of the twentieth century. After the earthquake of 1906, as Richter put it, "there was a feeling on the part of people like Lawson and Reid that it would be a good idea to detach this thing from the Weather Bureau and get it into more expert hands." By "this thing," Richter meant earthquake monitoring and investigations. And by "more expert hands," he meant scientists like Andrew Lawson and Harry Fielding Reid, geologists from the ranks of academia who had played key roles in post-1906 scientific investigations.

By Richter's account, the Survey was apparently not interested in the business of earthquake monitoring, but another agency did have a natural interest in earthquake exploration: the Coast and Geodetic Survey. Established in 1878, the organization's roots date back to 1807, when Congress directed that a survey of the coast should be carried out, and the U.S. Coast Survey was born. Geodesy is the science of determining the precise size and shape of the earth and typically involves repeated measurements at fixed "monument points." Such points were, by definition, assumed to represent a fixed point of reference from which coastlines and other geographic features could be measured. When the 1906 earthquake came along, some fixed points of reference were fixed no longer. During the earthquake, a large swath of northern California to the east of the San Andreas Fault slid south relative to the west side of the fault: some supposedly fixed reference points went north and some went south. The Coast and Geodetic Survey thus found itself responsible for data that could not simply be collected once and filed away forever. The basic data—the detailed positions of points on the earth's surface—could not be trusted to stay put.

The movement of survey points in fact represented one of the key scientific measures of the 1906 temblor: Reid used these data to piece together the process whereby strain (or energy) builds up in the crust prior to a large earthquake and then is released as a fault moves. Seismologists today know this as *elastic rebound theory*, one of the fundamental tenets of earthquake science. Elastic rebound theory leads naturally to the concept of an earthquake *cycle*: after an earthquake releases the strain on a fault, another earthquake cannot happen until sufficient strain has built up again. The theory of plate tectonics, developed in the mid—twentieth century, explains where this strain comes from. For a fault like the San Andreas, which represents the main boundary between the North America and the Pacific plates, the inexorable motion of the earth's plates provides the engine that drives the cycle.

By Richter's account, the Coast and Geodetic Survey employed a number of highly competent scientists in the early twentieth century, who stepped forward with interest in the earthquake problem. "Finally proper political stringers were pulled," Richter explained, "and action was put through Congress which officially transferred the responsibility for earthquakes from the Weather Bureau to the Coast and Geodetic Survey, where it remained until 1965. And quite a lot of work on earthquakes was then done by the staff and under the auspices of the Coast and Geodetic Survey."

The earliest U.S. earthquake monitoring program included the installation of seismological stations and the analysis of data collected. The CGS published a series of reports on damaging earthquakes and compiled documents pertaining to important historical earthquakes. They published a Quarterly Seismological Report during the years 1925–27 that included a compilation of shaking effects from earthquakes.

In the aftermath of the 1906 earthquake, scientists in California grew increasingly interested in setting up local laboratories to record earthquakes. The state's first seismological laboratory began operations at UC Berkeley in 1910 under the direction of the university's geology department. Harry Oscar Wood, who had surveyed damage in San Francisco in 1906 under Andrew Lawson's direction, joined the new lab. Wood had a background in geology and mineralogy and had completed some graduate school, but he had not attained his Ph.D.—a nontrivial point within the hierarchical halls of academia. Limitations associated with his lack of the proper credential sent Wood in search of a more congenial position. He went first to the Hawaii Volcano Observatory but soon began scheming to return to California.

Wood's role in the development of the Seismology Lab at Caltech, as well as the Seismo Lab's early history, is described in detail in Judith Goodstein's excellent book, *Millikan's School: A History of the California Institute of Technology*, and will be summarized only briefly here.

Wood's return to California was set into motion by two papers that he published in 1916, the second of which was titled "The Earthquake Problem in the Western United States." In this paper Wood laid out the blueprint for a network of seismic stations throughout California and made the strategic suggestion that the plan could be tested on a limited scale in Southern California. Wood knew, of course, that UC Berkeley already had the beginnings of a network, whereas Southern California was wide open. Wood was, moreover, convinced of the tremendous societal importance of the "earthquake problem" in both the southern and northern parts of the state. He knew that the last great earthquake on the southern San Andreas had been in 1857, compared with 1906 on the northern part of the fault. The most basic concept of an earthquake cycle suggested that Southern California represented the most likely locale for the next great San Andreas earthquake.

While all of this was going on, Caltech was busy getting itself off the ground: the institution essentially had no geology department until John Peter Buwalda was invited to head a graduate program in 1926. Wood, however, enticed the private Carnegie Institute in Washington to get involved. At the time, a Californian with a geology background, John Merriam, headed the Carnegie program. Merriam's interests may have paved the way for Carnegie's decision to pay the publication costs of the extensive "Lawson Report" that had been assembled under the leadership of Andrew Lawson to describe the 1906 earthquake.

In 1921 Carnegie agreed to fund a seismological laboratory in Southern California. The lab was set up under Harry Wood's direction, and operated in borrowed office space at the Mount Wilson Observatory on a mountaintop north of Pasadena. Wood recruited Richter as well as Hugo Benioff, a former astronomer, to join his new laboratory, and turned his own attentions to inventing a seismometer that could record California's local earthquakes. In this endeavor Wood joined forces with astronomer John Anderson, who had worked during the First World War on sensitive instruments

Fig. 5.1. Harry Oscar Wood (1879–1958). (Photo courtesy of Caltech Seismological Laboratory.)

to detect submarine vibrations. This collaboration bore fruit quickly, in the form of the Wood-Anderson torsion seismometer.

At a time when seismometers were heavy behemoths, the Wood-Anderson instrument was a marvel of engineering: reliable, compact, even portable. In Anderson's possibly immodest words, "There ain't no other seismograph worth talking about than ours." The first Wood-Anderson seismometers were running at Mount Wilson Observatory and in the basement of the physics building on the Caltech campus a year later. The instrument soon caught the eye of the head of the Berkeley operation, who ordered several for their own use.

Meanwhile, as Goodstein describes, the Pasadena Seismology Laboratory struggled to find its way in the world. The Carnegie Institute wanted to find a permanent laboratory, but was not willing to shoulder the cost of building a new facility. The head of the Mount Wilson Observatory, George Ellery Hale, proposed to Caltech that it provide a site and building so that future work by the lab could be carried out as an integral part of the university's geology and geophysics program. Negotiations ensued over the course of several years, with interest from Caltech but also concern on Carnegie's part over the possibility that it might lose independence in lab operations. A further concern arose in the person of Harry Wood: as an indispensable part of laboratory operations but lacking a Ph.D., he stood to lose out if his privately run laboratory became part of an academic department. An early agreement resulted in 1925: Carnegie agreed to pay for a laboratory to be built in the San Rafael hills in western Pasadena and research would remain under the exclusive auspices of the Carnegie Institute. The laboratory was first known simply as the Seismological Laboratory but was later named after the private foundation that helped finance its construction: the Kresge Lab.

The building itself was constructed with concrete walls thick enough to survive Armageddon; the arrangement between the two institutions, however, proved rather less enduring. The ties between the lab and Caltech grew stronger when the university hired Beno Gutenberg in 1930. Gutenberg had an office in the Seismological Laboratory, and began his close collaborations with Richter in particular. And thus began Caltech's sense of ownership in lab operations, as Caltech, not Carnegie, paid Gutenberg's not insubstantial salary. By 1934 Carnegie had made a decision to withdraw its support from the seismology program in Southern California. This move had been precipitated by Caltech's increasing insistence that since

Fig. 5.2. The original Kresge Laboratory, built by the Carnegie Institution in 1927, in the San Rafael hills of Pasadena. (Photo by author.)

they were paying the bills for the Seismo Lab, they should have some control over the program. In Goodstein's words, "After protracted negotiations, Millikan got what he wanted." Caltech formally took over operations of the Seismo Lab on January 1, 1937. Richter and Benioff became assistant professors. As feared, the transition left Wood out in the cold: he remained at the Seismo Lab as the sole employee on the Carnegie payroll.

As it happens, the organizational transition proved to be the least of Harry Wood's problems. In 1934 he contracted an incapacitating infection within his spinal cord. Wood was unable to work for several years; in 1937 he recovered well enough to return to the Seismo Lab on a part-time basis, but remained weak and prone to bouts of subsequent illness. Those who knew him in the 1940s and 1950s describe him as a bit of a hypochondriac: so vigilant about germs that he never joined the regular lunchtime bridge game at the lab, for example. (He had other reasons to avoid social gatherings as well, having been shy to the point of mortification around women.)

Harry Wood died in 1958, at the age of seventy-nine: a remarkably long life for someone with such serious health problems. With no closer relatives than a couple of distant cousins, he left his estate to the Carnegie Institution with directions that the money be used to fund grants to support studies focused on "geological aspects of seismology." Having apparently led an austere life, he left a substantial estate to Carnegie: bonds and stocks valued at over $400,000 in 1958 dollars. Among the members of the first Harry Oscar Wood Memorial Award Committee: his longtime colleague, Charles Richter. The committee first met in October of 1960, awarding its first two grants to Clarence Allen at Caltech and Inge Lehmann—one of the very earliest female pioneers in seismology—from Denmark.

Richter would later say that the field of seismology owed "a largely unacknowledged debt to the persistent efforts of Harry O. Wood for bringing about the seismological problem in southern California." Without question his nearly single-handed efforts launched the first seismological laboratory in Southern California. One might fairly say that earthquake monitoring in Southern California was inevitable given the interest generated by the 1906 earthquake—and the fact that post-earthquake investigations led geologists to map out nearly the full extent of the San Andreas Fault, including its southern half. Indeed, the writing was on the wall: Harry Wood was the right person in the right place at the right time to make it happen sooner rather than later. But without question, it would have happened not too much later, with or without Wood's efforts.

Intriguingly, however, Wood was responsible for another key development, one that proved critical to the business of early earthquake exploration in Southern California, and one that might never have happened at all if not for his efforts. It was Harry Wood who saw the need to recruit a young scientist with a background in physics; it was Harry Wood who took this appeal to Caltech president Robert Millikan; it was Robert Millikan who sought out the young Charles Richter and made a somewhat unwilling seismologist—but a seismologist nonetheless—out of this enormously troubled but also enormously talented young scientist. Take Harry Oscar Wood out of the equation, it is entirely possible you take Charles Richter out of the picture as well.

CHAPTER 6

The Kresge Era

> The laboratory routine, which involves a great deal of measurement, filing, and tabulation, is either my lifeline or my chief handicap, I hardly know which.
> —*Charles Francis Richter, 1949*

WHEN Charles Richter joined the Seismo Lab in 1927, it belonged to Caltech neither managerially nor physically. The laboratory had been built not on the central Caltech campus, but rather in the gently rolling hills of western Pasadena. Known later as the Kresge Lab, the facility continued to house Seismo Lab operations until the group moved to a new building on the Caltech campus in 1974. Kresge, outwardly a large, homey two-story building with a Spanish façade, offered a number of attractive amenities to early seismologists. Notable among these were the quiet location away from central Pasadena and the hills themselves, solid granite into which a vault could be constructed. This vault, essentially a man-made cave, was carved into the hillside against which the main lab building was situated, providing an especially quiet setting for sensitive seismometers. The seismometers furthermore rested on "piers"—blocks of concrete built in such a way as to be directly connected to the ground but not to the foundation of the laboratory building. Piers are anchored directly into the ground, in this case solid bedrock: the foundation and the structure are built around the piers. Thus the recordings of earthquakes remain relatively uncontaminated by the shaking of the building itself. The solid granite underpinnings, meanwhile, guaranteed that minimal cultural noise, such as automobile traffic, would interfere with the recording of earthquakes.

The new laboratory building had been completed and instrumented in early 1927, just months before Richter joined the staff as a young assistant. In addition to the seismological amenities, it offered other comforts: an

Fig. 6.1. Structure of the Kresge Lab. The back and attached tunnels were carved into the solid rock of the hillside. (Photo by author.)

Fig. 6.2. David Johnson behind one of the isolated piers built to allow seismometers to record signals from the earth without contamination from vibrations of the structure. Note that the floor is not connected to the base of the pier. (Photo by author.)

Fig. 6.3. The Kresge Lab in 1929. (Photo by James B. Macelwane, from James B. Macelwane Papers, St. Louis University archives.)

outwardly lovely building in a quite, wooded setting; occasional deer ambling through the grounds. When the Carnegie Institute set out to build a seismological laboratory, they had in mind, essentially, a bunker: a spartan, utilitarian lab space of heavy-duty concrete construction where the business of science and instrumental design—including the recording of earthquakes on sensitive instruments—could be undertaken in peace and quiet. The Kresge Lab was to be built, however, in the residential and affluent San Rafael hills. For officials on local zoning boards, a bunker wouldn't do. Thus was the inwardly Spartan Kresge Laboratory built with outwardly Tuscan trappings: especially if one didn't look too closely, from the outside the building could be mistaken for a lovely two-story residence. The impression quickly dissipated as one climbed the stairs and walked through the front door, although the large entry room had some of the comforts of home, including a fireplace.

By the account of Hertha Gutenberg, wife to Richter's longtime colleague and collaborator Beno Gutenberg, "They all loved the old lab." She continued, "I know people who were there for a long time, and when they all came together they would say, 'It's not the old Seismo Lab anymore.' They really were a close group." Betty Shor, daughter of James A. Noble,

Fig. 6.4. Mildred Lent, Hugo Benioff, and Charles Richter at the Kresge Lab. (Photo courtesy of Caltech Seismological Laboratory, reprinted with permission.)

Caltech professor of mining geology, wife of Richter's graduate student George Shor, and, eventually, Richter's paid assistant, describes a congenial routine in the 1940s and early 1950s: gatherings over coffee (never snacks, just coffee) and conversation every morning at ten and every afternoon at two or two-thirty. Everyone—scientists, students, and staff—usually stopped by. At noon Richter and others often gathered at a central table for a game of bridge. The tradition of morning coffee continued through the 1960s, by which time lab operations had expanded into a former private residence across the street. Electronics technician Bob Taylor, first hired to work on the seismometer that went to the moon, describes coffee breaks in the heating room in the basement: chairs on one side of the room, heating and air-conditioning ducts on the other. Taylor recalls the room as a dungeon, but still expresses palpable affection for the gatherings it housed.

From the time the lab began, the design and construction of new and better seismometers occupied the attentions of many at the lab, scientists and engineers alike. The field of "seismometry" remained very much in its infancy in the early years of the twentieth century: earlier seismometers were large, heavy, expensive, and designed to record large earthquakes worldwide rather than smaller earthquakes close by. Even today, scientists generally rely on different types of seismometers to record global versus local and regional shocks, although the newest (and most expensive) modern seismometers are able to serve both purposes. It remains now, as it was in the early days of the Seismo Lab, no easy matter to design a seismometer: the business of seismometry required considered scientific acumen as well as engineering prowess.

During the early years of the laboratory, Harry Wood had, as we have heard, teamed up with John Anderson to develop the so-called Wood-Anderson torsion seismometer, an instrument that quickly set the standard for recording of local earthquakes. In 1926 the laboratory began a cooperative project with the geology division of Caltech to install the first network of earthquake-monitoring stations around Southern California. Today a small team of scientists and technicians can deploy a small handful of seismometers in a day. In the early decades of the twentieth century, earthquake monitoring was a very different ballgame: it would be six years before Wood and his staff had a network of six instruments in place and had begun to use the records from these instruments to produce a regular bulletin.

When earthquake scientists turn to the Web to download information from the earthquake data in Southern California, the catalog they find begins in 1932. From this date, earthquakes in the region were cataloged routinely, their basic parameters—including magnitude—quantified and preserved in a consistent format. The network, analysis methods, and catalog format have evolved, but the catalog is made retrospectively consistent to the extent possible. This is no mean feat: the basic data from which earthquake catalogs are derived have evolved from five-inch by seven-inch "phase cards"—index cards, basically—upon which raw readings were penciled to the sophisticated digital data of the modern computer age. Efforts to reinterpret old data with more modern methods continue to this day.

At Kresge Richter found himself a part of a new endeavor in earthquake science: a laboratory dedicated to the business of recording and analyzing earthquakes. He did not contribute to seismometer design efforts: even those who best appreciated Richter's scientific talents are quick to acknowledge that he was no genius with mechanical devices. (We are, one might recall, talking about a man who had left the field of chemistry because of "unfortunate incidents" in student labs.) Instead, Richter focused on analysis of data from the new instruments, the complicated wiggles etched onto film that provided some of the earliest instrumental recordings of local earthquakes.

In the analysis of seismograms as with the design of seismometers, it fell to Richter and his colleagues to pave the way for future generations of seismologists—so new was the entire business of earthquake monitoring at the time. Only a handful of such labs existed in the word, the largest in the United States being at UC Berkeley, Lick Observatory east of San Jose, and St. Louis University. St. Louis University, like a number of other laboratories elsewhere in the country, was a Jesuit institution. A religious order well known for its tradition of scholarship, the Jesuits have played key roles in the sciences for centuries. By 1750, 30 of the world's 130 astronomical observatories were run by Jesuit astronomers. Jesuits would play an even bigger role in the early days of earthquake science. Seismology has even been called the "Jesuit science," so important were their efforts to the early days of seismology in the United States.

Jesuit Father James B. Macelwane was a driving force behind the laboratory at St. Louis. According to Richter, "he used his connection with the church to further the development of seismology at various Jesuit institutions." Richter explained further, "It was said that this got along well because this was a scientific development and no one could see any way that it could come in conflict with the church dogma. Whatever the circumstances, this was so." The Jesuits began operating seismological stations in 1868 in the Philippines and by 1916 operated several dozen stations in locations as far-flung as Cuba, Tarragona, China, Australia, La Paz, and Milwaukee.

Father Macelwane was a remarkable figure in his own right. Ordained into the priesthood in 1918 at age thirty-five, he received a Ph.D. in physics from the University of California in 1923. He went on to conduct seminal research into the deep structure of the earth, and in 1936 published *Geodynamics*, part 1 of *Introduction to Theoretical Seismology*. It was said to be the first book with a detailed presentation of seismic wave theory that could be understood by a beginning student. In 1947 he published a popular book, *When the Earth Quakes*, to explain earthquake science to a nonspecialist audience. He served tirelessly on committees, often assuming leadership roles in organizations such as the American Geophysical Union. Interestingly, it was under his tutelage that Florence Robertson became the first American woman to receive a Ph.D. in geophysics, in 1945. The AGU today awards the James B. Macelwane Medal annually for outstanding scientific contributions by young scientists.

The research group that Macelwane established at St. Louis University remains one of the leading centers of seismology research in the United States. Over time, however, there has been a steady westward migration of the center of mass of earthquake research in the United States. California, and the West Coast as a whole, are where the action is, earthquake-wise; so too have they gotten to be where the action is, earthquake-science-wise. This preeminence traces its roots back to the pioneering efforts in northern and southern parts of the state; to the network that began in the latter region in 1932 and has continued to grow—not steadily but rather in fits and starts—since that time.

The early Seismo Lab grew and then divided itself in 1957. The Kresge Lab itself was not a small structure, with two full floors as well as a modest maze of tunnels and piers designed to house instruments. As the Seismo Lab operations grew, however, the quarters seemed to shrink, bursting at the seams with scientists' offices, laboratory space for instrument development and design, analysis rooms, and an ever-increasing need for storage to house the ever-expanding collection of data. Wooden storage cabinets had even been built under the attic eaves—and will likely remain there until the end of time because they are very nearly part of the building. (It would take a mighty effort to demolish the lab building itself, but its historic status may preclude this in any case.)

Its occupants eventually pressed for space, Caltech considered an addition to the structure, but opted in the end for what was, paradoxically, a cheaper solution than adding onto a building with massively thick concrete walls: they bought a private residence across the street from the lab, what became known as the Donnelley Lab, when it came on the market. The acquisition required negotiation: a zoning variation was necessary to use the former house as office space and the nearest neighbor objected vociferously. By Richter's account, the man had been angling for a generous buyout of his own property, but in the end Caltech prevailed and the move commenced. Thus began an era of stratification: the scientists, including Gutenberg, Richter, and Wood, moved up to the "mansion" up the hill, while technicians and engineers remained at the old lab. The new digs offered not only more space but also an amenity after a seismologist's heart: an underground elevator shaft and tunnel dug one hundred feet into solid granite—a perfect additional testing grounds for one particular type of instrument under development at that time.

Seismo Lab seismologist Hugo Benioff, who made seminal contributions to both seismometry and seismology throughout the course of his long career, used the elevator shaft to house instruments known as strainmeters. And herein lies the beginning of, not a technical digression, but a brief R-rated anecdote from the Kresge era. Strainmeters measure not the motion of the ground during earthquakes, but the minute warping of the earth

Fig. 6.5. Private residence that housed Seismo Lab operations between 1957 and 1974. (Photo by author.)

caused by strain, or energy, from plate tectonic (or other) forces. Such instruments are, even today, difficult and expensive to build, and difficult to operate, as various types of cultural noise can easily overwhelm the tiny signals one hopes to measure. (The earth's crust does not in fact ever warp by very much.) Benioff's strainmeters proved useless during the day, when the ever-present hum of traffic and lab operations drowned out strain signals from the earth itself. At night, when the lab and its environs grew more quiet, the instruments had their chance to collect useful data. Benioff noticed, however, that unexpectedly high levels of noise appeared commonly during nighttime hours. He and his colleagues eventually came to understand the origin of these signals: the long, quiet drive leading up to the lab had become a local lovers' lane. A junior scientist suggested adding a sensor to the instrument that would register telltale nocturnal vibrations and switch on floodlights along the path. The lab instead decorously opted to install a chain across the entrance to the driveway.

Returning to our G-rated discussion, in August 1957 Richter wrote a letter to the director of the Seismological Laboratory, Frank Press, to report on the move. Richter talked about "initial confusion" and "some nerve strain" over the overall disruption as well as petty annoyances such as difficulties with phone lines and a shortage of office furniture. Still, he noted, "I think most of us are very happy with the new arrangement." Perhaps unbeknownst to Richter, Press had in fact worked behind the scenes to see that Richter was given adequate office space in the new lab.

Even as it expanded, the Seismo Lab remained a small and elite fraternity. No more than about a dozen worked on the technical staff at any one time, with even fewer research scientists in residence. A spattering of secretaries and a small stream of graduate students rounded out the rolls. Those who were a part of the era remember it with fondness. David Johnson, later a senior instrumentation specialist with the Seismo Lab, joined the Kresge Lab in 1965 in a capacity that he never dreamed would be permanent. Johnson recalls the Kresge era with nostalgia: the remarkable people of the era in particular, starting with Francis Lehner, then head of the laboratory operations at Kresge. A perfectionist and old-school engineer who towards the end of his career had no use for newfangled transistors, Lehner helped design many of the early seismometers built at the lab, including the Ranger seismometer sent to the moon. In 1965 Lehner offered Johnson, whom he knew through a family connection, a weekend job changing and developing seismographic records.

At the time, the seismic network comprised nine stations, and the routine business of developing the records was a long way from the automated

Fig. 6.6. Hugo Benioff with one of the instruments ("strainmeters") that he designed and operated at the Kresge lab. (Photo courtesy of Caltech Seismological Laboratory, reprinted with permission.)

procedures that networks employ today. At that time, seismometers recorded data on film that had to be kept away from light—or under the eerie red glow of a "safe light" until it could be developed. Film records from stations elsewhere in California were sent in, in special black envelopes, once a week by mail, mostly by a network of volunteers who had been enlisted to help with early—and very poorly funded—earthquake-monitoring efforts. The film that arrived by mail had to be developed,

Fig. 6.7. Francis Lehner, Seismo Lab engineer. (Photo courtesy of Caltech Seismological Laboratory, reprinted with permission.)

along with film from the instrument at the lab itself. The recording film had to be changed daily in the seismometer at the Kresge Lab, and the instruments as a rule demanded considerable babysitting.

During the week some key routine operations were handled by Seismo Lab staff, including, by the mid-1950s, two women: Violet Taylor, known to her friends as Vi, and Gertrude Killeen. A third woman, Phyllis Cangelosi, was described by Richter in 1957 as "developing into a valuable draftswoman." It is Vi Taylor, however, who emerges as the real (female) fixture

Fig. 6.8. Violet Taylor, circa 1955 (left). (Photo courtesy of Caltech Seismological Laboratory, reprinted with permission.) Gertrude Killeen, circa 1955 (right). (Photo courtesy of Caltech Seismological Laboratory, reprinted with permission.)

at the lab for many years. A sharp, lively, and energetic woman who brooked no nonsense, Vi handled routine processing of seismograms and trained many generations of graduate students whose duties involved assistance with lab operations. Those who watched the lab in operation, for example graduate student Shelton Alexander, soon realized that Vi essentially kept the so-called measurement room running, her considerable managerial skills making up for Richter's (limited) abilities in that realm. Another graduate student, David Hill, gives a nod to Vi's talents as well: Vi managed keep her young charges in line (and sometimes her boss) and yet at the same time create an enjoyable working atmosphere for them. For a time the grad students' duties included handling routine processing tasks on the weekends, but by the mid-1960s some students had begun to balk at the arrangement. Johnson accepted the offer to fill in, attracted by a paycheck that would supplement his earnings from his regular job with his family business. He continued to work only weekends for many years before taking on more substantial duties maintaining the ever-expanding seismic network.

Johnson describes a congenial atmosphere at the lab, even after the scientists moved to the house up the hill. "We had great parties," he said, "at the house and sometimes outside." By the mid-1960s the lab operations kept about a dozen technicians and engineers employed. There were female secretaries over the years who tended to come and go quickly, often for the same reason—in that era, secretaries often stayed employed only until a first baby was on the way. Vi and Gertrude, however, remained for many years, both very much a part of the fabric of the lab. A chain-smoking Irishwoman, Gertrude also brooked no nonsense—with anyone. She made coffee for the lab, and, in Johnson's words, "you didn't complain about the coffee." He further hints that one *might* have complained about the coffee: Gertrude's measuring technique, at least where coffee was concerned, was reportedly haphazard at best. Betty Shor recalled Gertrude as having done an outstanding job printing the photographic records, adding that "one could tell when things weren't going as well as she wished, because she would burst into some great swearing."

Although perhaps not exactly your typical woman of the 1960s, Gertrude Killeen was stereotypically feminine in one respect: her visceral dislike of lizards. Although the rough-and-tumble atmosphere of the lab suited her personality in many respects (one cannot quite picture her as a schoolteacher or a nurse, let alone a nun), the Kresge Lab setting was not the best place to be if one took exception to reptiles. The semiarid hills of Southern California were, and to a lesser extent still are, literally crawling with small desert lizards. That they are entirely harmless came as no consolation to Gertrude. Her job responsibilities required her to walk up and down a long, winding stairway up the hillside just west of the Kresge Lab—the shortest route between the lower lab and the upper Donnelley House. If she got to the bottom of the stairs and found a lizard on the ground sunning itself, as lizards are wont to do, she would turn around, walk all the way back up, and take the much longer path along the road and driveway. Within the lab, occasional ear-splitting shrieks from the women's bathroom meant only one thing: a hapless lizard had found its way inside and invaded Gertrude's personal airspace.

Vi, for her part, reserved her visceral dislike for snakes—another unfortunate trait for anyone working in the scrubby California foothills. She was, however, quite fond of local foxes, setting bowls of milk out for them every morning. Such stereotypical feminine traits aside, Vi and Gertrude worked shoulder to shoulder with the men in the lab, contributing to the day-to-day operations in a way that was scarcely common at the time. Richter moreover very much appreciated Vi's contributions to the lab. In 1961 he wrote Frank Press, arguing that a raise of her salary was in order. Richter wrote further that Violet was very nearly capable of running the

Fig. 6.9. Hillside between the lower Kresge Lab and the upper Donnelly House. (Photo by author.)

lab operation in his absence and suggested that a change in job title would be appropriate, to "seismologist." His letter also praised the efforts of his secretary and suggested that her salary should also be raised, but if funds were limited, not at the expense of Vi's raise.

Richter's support for the female staff members at the Seismo Lab appears to have been unwavering: from the recognition and praise of Vi's contribution at a time when such contributions from "untrained" women in science were often denigrated, to his respectful working relationship with Betty Shor, to the support he showed for the young women who came and went in a secretarial capacity. When Betty Peach left her secretarial position in 1963 (indeed, because she was expecting a baby), Richter took the time to sing her praises again in a letter to Press, and expressed regret at her departure.

Richter's respect for his female colleagues may have been a natural consequence of his general respect for talent. Throughout his life he revealed an unerring eye for, and appreciation of, capable people. He took the time to applaud in writing not only promising students and scientists but also journalists and others who had impressed him. For anyone who has had the pleasure of meeting Vi Taylor, Betty Shor, and Gertrude Killeen, there is no question that they were remarkably talented and capable individuals. Shor herself points laughingly to another motivation: Richter found himself in need of an assistant, and "who was cheaper [to hire] than a grad student's wife?" Clearly, Richter saw in her far more than cheap labor, and happily put her talents to use.

The Seismo Lab's women contributed to the operation in a variety of ways: routine analysis, film processing, figure drafting, as well as secretarial work. The larger number of men were engaged in a number of principal activities in addition to research. Graduate students—all of them male through the Kresge era—arrived to learn from preeminent scientists in their field, many of them going on to eminent careers in other institutions, and in a few cases at Caltech. During Richter's tenure, twenty-six students completed Ph.D. degrees, several of whom (including Clarence Allen, Don Anderson, and David Harkrider) remained at Caltech as professors. Thomas Hanks, who arrived as a student in 1966, talks with open affection about his time at the lab, and in general for an era when "hands-on science meant something other than having your hands on a computer keyboard." Seismo Lab students from that era learned seismology from the ground up, literally. Even if they were not involved with instrument design, they rotated in one-week stints as assistants with the most basic routines and analysis, essentially assisting Vi Taylor once they had come up to speed on procedures.

Nowadays a great deal of basic data processing and analysis is done by computers. At the Seismo Lab today, the additional routine processing that must still be done by humans is handled by a team of capable analysts as a part of the responsibilities of their full-time jobs. In Tom Hanks's day, it fell to students, under Vi's watchful eye, to take records of seismograms (by that time paper rather than film), and read the times of P and S waves as well as other critical observations, such as the amplitude of the peak shaking that was required for a magnitude calculation. (See sidebar on p. 78.) Results ended up hand-written on the five-by-seven phase cards, great stacks of which remain in storage today. As late as the opening years of the twenty-first century, lab analysts, under the supervision of lead lab seismologist Kate Hutton, continued to enter data from these cards into computer files. Once the readings were computerized, the earthquakes could be located using far more sophisticated methods than had been available in earlier times.

Even today, a great deal of seismology analysis begins with basic measurements that were familiar in Richter's day: the amplitudes of waves on a seismogram and the precise time at which waves arrive at any given station. When a fault moves in the abrupt lurch known as an earthquake, the process sends energy careening into surrounding earth in all directions. This energy takes the form of different wave types. The fastest of these is the P, or primary, wave, a compressional wave akin to the disturbance one can generate by stretching a Slinky toy taut, plucking a coil back, and letting it go. The S, or secondary wave, travels more slowly; this wave in the earth is like the disturbance created by pulling part of the Slinky sideways relative to its length and then letting it go, creating the familiar sinuous wave ripple. Conveniently, the letters P and S also stand for the physical nature of each type of wave: pressure and shear. Other, more complicated waves are generated as well, including the so-called surface waves, which ripple along the earth's surface and, if the earthquake is large, can ripple around the world many times over. Surface waves are now routinely recorded at large distances (thousands, even tens of thousands, of kilometers) by specialized instruments deployed around the globe; as a rule they are not perceived by humans. When a local or regional earthquake strikes, observers typically feel the initial P wave followed by the S wave, which moves at a more languid pace but is almost always stronger.

Hanks describes a congenial atmosphere within the Seismo Lab, with no discernible stratification between the professors and students who worked in the upper Donnelley Lab and the staff of lower Kresge Lab. He recalls fond memories of men like Francis Lehner, who eventually married Vi Taylor several years after the death of his first wife.

Within the Donnelley Lab itself, Hanks recalls Richter as having been mostly "around, doing his own thing." By that time Richter was in his late sixties, perhaps even less inclined towards interactions with students than he had been in earlier years. Hanks may not have been aware of Richter's interactions with one particular member of the Kresge Lab staff: Bill Gile. By Bob Taylor's account Gile was every bit as opinionated as Richter: the two would get into long, heated, and loud arguments, then retreat to their separate offices to fume in private. But the separations never lasted long: within hours or days they would be back in one office or the other, arguing about something else. "They both loved it," Taylor observes with a grin.

Hanks describes Richter more benignly: a "typical, moderately eccentric scientist." It would perhaps be fair to say that Caltech has, and has always had, its fair share of such individuals. Richter did distinguish himself in one respect: by Hanks's recollection, he always wore a coat and tie, even when most of the men around him were in "shorts and T-shirts." The latter was surely hyperbole, but even today most scientists at the Seismo Lab and elsewhere are not generally given to formal attire. When a large earthquake struck Southern California in the wee hours of a Saturday morning in 1999, many Seismo Lab and U.S. Geological Survey scientists appeared on television looking perhaps even more informal than their usual informal selves. The casual atmosphere did not escape the attentions of the public: one television viewer wrote to the *Los Angeles Times* to ask why "earthquakes always happened on dress-down day at Caltech." In his choice of attire, Richter was without question a breed apart during his later years—yet the impressive array of stain spots on almost all of his ties perhaps redeemed him from being a cut above.

A persistent tendency towards rumpledness also detracted from an impression of well-groomed attire. By the end of the day his shirttail was no longer reliably tucked in, nor his fly reliably zipped. Shelton Alexander describes the day that his wife, Judy, spied Richter on a local bus: having only met him once before, she recognized him by his disheveled attire as much as his looks. By the late 1970s Richter had become so permanently rumpled that Seismo Lab secretary Dee Page, her own grooming, attire, and comportment unfailingly impeccable, found herself constantly stifling the urge to smooth Richter's collar and straighten his tie.

Another graduate student from the same era as Hanks, Tom Jordan, began his graduate studies in 1967, having been employed previously by lab scientists Jim Brune and Clarence Allen to run a field experiment. Jordan echoes Hanks's recollections, describing Richter as "not easily approached," very much inclined to keep to himself. Richter would frequently go outside Professor Don Anderson's office to "bang the Wang." Anderson had invested several thousand dollars to buy one of the earliest commercial calculators, built by Wang Laboratories, that could compute a logarithm with a single keystroke. The functions of this machine could later be handled by the HP-35 calculator, which retailed for the low price of $395 in 1972. Today's HP30S handheld calculator will set one back about $20, with capabilities that far surpass the Wang that Richter banged.

The purpose of all of this banging was often something other than seismology. As early as the 1940s, Richter devoted a great deal of time and effort to enormously tedious number theory calculations, in particular focused on determination of large prime numbers—his idea of an amusing diversion to occupy idle moments. Among Richter's papers at Caltech there is one box full of nothing but such calculations, done laboriously by hand.

Some older scientists resist new technologies when they appear, but not Charles Richter: when Anderson's Wang calculator appeared, he pounced on the opportunity to speed up his calculations.

Hiroo Kanamori first arrived at the Kresge Lab as a postdoc in 1965, by which time even the "mansion" was pressed for space. Kanamori's first "office" was a small desk tucked into a corner of a large conference room. A consummate diplomat as well as scientist, Kanamori would later credit his central location for having provided him with abundant opportunities to interact with faculty, students, and visitors. When he later moved to new quarters, his office provided a different opportunity for interaction: during his first few months, "many visitors inadvertently rushed into [my] office." He eventually learned that to make space for his new office, the Lab had converted a bathroom that had been attached to Hugo Benioff's office.

By the time the lab reached the half-century mark, it had come a long way since its shoestring inception. Carnegie's Seismo Lab became Caltech's Seismo Lab, and Caltech's Seismo Lab had grown enormously in size and stature. Yet over its first five decades lab operations remained housed in the quaint building tucked among the live oak trees in the rolling, lovely hills of west Pasadena—and in the mansion next door. Eventually, however, the central campus beckoned, offering more space as well as closer interactions with the rest of the geologic division. Following the Seismo Lab's move to the main campus in 1974, the Donnelley Lab was sold, reverting back to its original function as a large, elegant home in the San Rafael hills. Tom Jordan described the unnerving experience of joining his wife on a Pasadena home tour in the year 2000, after he had moved back to the Los Angeles area to become head of the Southern California Earthquake Center at the University of Southern California. To his surprise, one of the homes on the tour was none other than the old lab, by then decked out with all of the requisite accouterments of a house in the upscale neighborhood. Finding himself in a bedroom full of chintz and other finery, Jordan informed his tour guide that they were standing in Charlie Richter's old office. The man needed a few seconds to digest this bit of seemingly incongruous information: he had no idea he was standing in a part of Southern California history.

The Kresge Lab itself continued to house data, instruments, and occasional lab experiments conducted by other Caltech scientists. For many years, not only the abundant built-in storage shelves but also much of the lab equipment remained, giving the vacated rooms and hallways a ghostly feel. Even the red "safe light" still glowed at the flip of a switch in the room where seismometers of old once etched their data onto photographic film.

Caltech staff and visitors would sometimes make their way over to the lab, often on official business—most frequently tending to the current incarnation of the seismic network station, known simply by its three-letter station code, PAS, in earthquake circles—that has since the very beginning provided a key "reference" station for the Southern California network. (Because the instruments in the vault are on such solid rock, their data is of particular importance for many types of research.) Sometimes, however, people would come for other reasons, drawn by the charm of the setting—the nostalgia of a bygone era. Sadly that era came to a close in 2005, when Caltech relinquished ownership of the quaint but largely unused property. The land ("Almost 3-acre gated compound.") was sold to a real estate developer with plans to build luxury homes on the property. The network station, PAS, was taken out after over seventy years of faithful operation, replaced with a new site not far away in the hills of northwest Pasadena.

In a monumental and exhausting project, David Johnson and his colleagues oversaw the disposition of the lab's contents: furniture, equipment, and a veritable mountain of old but potentially valuable data. Some of the old records could be thrown away, in particular the large percentage of film records that had been saved even though they contained no earthquake recordings. Other data migrated to a basement storage facility in another piece of property that Caltech had acquired for office space. This data did not, of course, migrate of its own volition. One is inclined to be impressed that David Johnson does not lose his usual twinkle and smile as he relates the story of cleaning out the house, in particular the data from the shelves tucked under the attic eaves, which had to be hauled down a steep ladder. One is also inclined to hope that he received hazardous duty pay.

Perhaps the ghosts will wander away from Kresge Lab as well now that the last vestiges of its glory days are gone. For anyone who remembers those days as well as those who have only heard the stories, those heady days remain firmly etched in not only their memories but also in their hearts.

CHAPTER 7

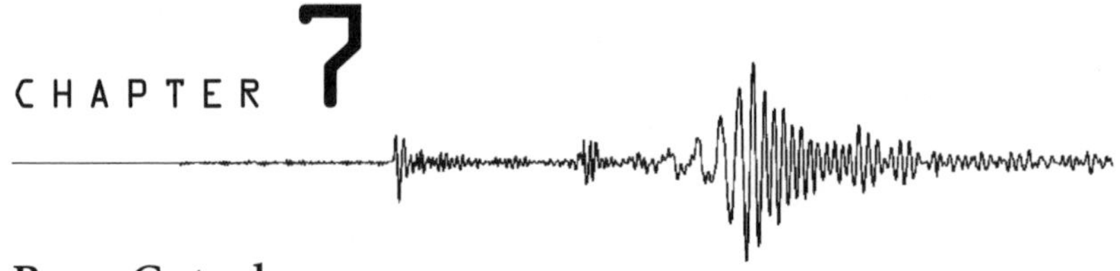

Beno Gutenberg

I owe a very great deal to him and came to regard him with almost a filial affection. I was very, very fond of that man.
—*Charles Francis Richter, 1979*

The Seismo Lab might have been up and running in 1927 when Richter arrived, somewhat ignobly, as a twenty-seven-year old assistant, but the institution remained very much in its infancy. Richter had already been on the staff for two years when Caltech evaluated the progress in the young program and planned a conference with the intention of hiring their own seismologist. The meeting organizers invited a number of luminaries from around the world, including Harry Fielding Reid, Father Macelwane, Perry Byerly from Berkeley, Harold Jeffreys from England, and Beno Gutenberg from Germany. None of these names have become household words, yet all are recognized universally in Earth science circles as founding fathers of modern earthquake science. Perhaps unfortunately, if Jeffreys's name is ever mentioned outside of narrow scientific circles, it is often to note that he was one of strongest opponents of plate tectonics when it emerged on the scene just after the middle of the twentieth century.

But I digress. The Seismo Lab staff, including Richter, planned an impressive event for their distinguished visitors. Nowadays the typical scientific workshop might be a few days long, or as much as a week. International meetings, to which participants travel from sometimes great distance, can stretch to two weeks, but few scientific meetings are this long. In 1929, however, international travel was no easy feat, and so the Seismo Lab planned an event to make the arduous trip worthwhile. The local media took note of the auspicious occasion. On October 3, 1929, the front page of the *Los Angeles Times* featured a banner headline, "They Study Ma

Fig. 7.1. Attendees at a 1929 workshop held in Pasadena. Among the luminaries in this photo: (*lower row*) (*3rd from left*) Hugo Benioff, (*4th from left*) Beno Gutenberg, (*5th from left*) Harold Jeffreys, (*6th from left*) Charles Richter, (*3rd from right*) Harry Wood, (*far right*) John P. Buwalda, (*top row*) (*2nd from left*) Perry Byerly, (*far right*) James B. Macelwane. (Photo courtesy of Caltech Seismological Laboratory, reprinted with permission.)

Earth's many Shivers and Moans," and a photo of workshop participants with the caption, "Famed experts will discuss disturbances over world."

The workshop began on October 2 and ended on October 16, with many days of scientific presentations and three substantial excursions: one to the San Andreas Fault north and northeast of Pasadena, one to Santa Barbara, and one up to the operations at Mt. Wilson Observatory. Photographs from the first excursion show a number of sites familiar to the present-day Earth scientist in the Southern California region, except that the fault features are revealed without obscuration by later development.

For most visitors to Southern California an excursion to the San Andreas Fault would possibly not rank high on the list of must-see attractions.

North of the Los Angeles region the fault skirts the northern flank of the San Gabriel Mountains, running adjacent to Palmdale and through small mountain towns such as Frazier Park, Gorman, and Wrightwood. The scenery is sometimes rugged, sometimes arid, and to a large extent, dull as dirt to the average tourist, except perhaps during ski season. The fault itself would moreover probably disappoint the casual observer, revealing no dramatic, fresh scars or gaping chasms across the landscape but rather large-scale topographic features, ridges and valleys far more easily appreciated from the air (not quite practical in 1929) than from ground level.

Faults like the San Andreas beckon to earth scientists as flowers call to bees. Just a year prior to 1928 a meeting had been held in northern California, the Marin Peninsula Sixth Pacific Science Congress. This meeting also featured a field trip to the San Andreas, specifically to visit the still-fresh scar left by the great 1906 San Francisco earthquake. This workshop was attended by some of the most accomplished scientists of the day, including Beno Gutenberg; the modern earth scientist is inclined to suspect the field trip rather than the workshop itself provided the allure.

Indeed, the Seismo Lab staff knew exactly what they were doing when they planned their workshop, including their excursions. To a scientist whose foremost passion is earthquakes, one could scarcely imagine more attractive amenities: not only a growing collection of state-of-the-art data, not only abundant local earthquakes, but also one of the most spectacular field laboratories anywhere in the world for the investigation of earthquakes and faults. The climate wasn't half-bad, either.

"It was commonly understood," Richter wrote, "among the whole Pasadena group that in all probability one of our distinguished foreign visitors would be invited to come to us, either on a temporary or permanent basis." By "foreign visitors" Richter meant Jeffreys and Gutenberg, who hailed respectively from England and Germany. What Richter perhaps did not know was that a decision had been made at high levels within Caltech, by then starting to feel a sense of ownership of the Carnegie-run lab. The new, state-of-the-art seismometers were starting to generate large volumes of data, data that could and should be put to use to address seismological and geophysical problems of fundamental importance. To reach its full potential, however, the Seismo Lab required a new infusion of expertise. Millikan turned his attention to Europe, where many of the brightest stars in the young field resided. Millikan and his contemporaries considered extending an offer to Byerly and Jeffreys, but had set their sights on another target, Beno Gutenberg, even before the workshop.

Fig. 7.2. Marin Peninsula Sixth Pacific Congress in 1928. Meeting attendees included (*3rd from right*) James B. Macelwane and (*2nd from right*) Beno Gutenberg. (Photo from James B. Macelwane Papers, St. Louis University Archives, used with permission.)

Gutenberg apparently impressed Millikan enormously during the conference. As Gutenberg departed, Millikan remarked that he hoped he would be seeing him again soon. Gutenberg asked in return, "Oh, are you coming to Germany?" Millikan did not reply; Beno and his wife Hertha would later conclude that the die had been cast. Hearing a rumor that Harvard was also interested in wooing Gutenberg from Germany (in fact Harvard had offered him a position two years earlier), Millikan sprang into action, sending a telegram to Gutenberg: "Could you consider seismological post here if satisfactory arrangements could be made?" The telegram arrived at the Guttenberg household when Beno was away at the university, and came with a prepaid reply. Hertha called her husband and asked if she should "telegraph back, 'Thank you very much, but no'?" Her husband told the director of the institute about it, who said, "Better go home before Hertha does something foolish." In fact, of course, Hertha had only been joking. Guttenberg later wired back, "Would consider post if arrangements satisfactory."

Fig. 7.3. Beno Gutenberg (1889–1960). (Photo courtesy of Caltech Seismological Laboratory, reprinted with permission.)

A formal appointment offer was sent to Gutenberg in January 1930, offering him a full professorship in geophysics and meteorology at Caltech. True to his word, Gutenberg accepted, but not straightaway. A half-hearted attempt was launched belatedly to find Gutenberg a suitable position within Germany. Some correspondence with Caltech ensued, a dialog that lasted several months. (International conversations were, of course, very much slower in those days.) Gutenberg suggested that the initial salary offer of five thousand dollars was too low; in the ensuing negotiations Millikan described the low cost of living in California and the relative cost of household help in the United States versus Germany. Millikan talked about the house in which he was then living: in one of the best quarters of town, on an 86 by 275 feet lot, with four bedrooms, a sleeping porch, and maid's quarters. The house was at the market at the time for twenty-three thousand dollars. (That same house might today fetch a hundred times more, give or take.)

Gutenberg persisted in his negotiations. By late May Millikan had raised his initial salary offer by 40 percent, to seven thousand dollars—a salary that put that put Gutenberg in elite company at the growing university, and that was substantially higher than Richter's. Gutenberg accepted the offer in June 1930, and by September arrived in Pasadena with Hertha and their children, Arthur and Stephanie. Gutenberg was not, Richter notes, a member of the laboratory staff per se, but was given an office at Kresge and spent a good deal of time in the lab.

The salary negotiations notwithstanding, it had not been terribly difficult to woo Gutenberg away from his native Germany. By 1930 Gutenberg was forty-one years old and one of the most highly respected seismologists in the world. His position in Frankfurt carried an impressive title—Professor Extraordinarius, but a far less impressive stipend, one that could not support his wife and children without assistance from his extended family, who operated a soap factory. By Richter's account Gutenberg's position was largely honorary, although he did earn additional income from editing and writing, and had a couple of technical books in print.

As it happens, Gutenberg had other reasons besides money to move halfway around the world. "After all," Richter explained, "he was Jewish, and there were already indications of trouble. After he was over here he went to considerable trouble and expense to help other people to get out of Germany before the storm broke." A scientist of diminutive stature but towering reputation, Gutenberg would have been an obvious choice to succeed pioneering seismologist Emil Wiechert at the seismological laboratory in Göttingen or to fill an open position at Potsdam. But neither offer was forthcoming, and, according to geophysicist Leon Knopoff's memorial, "these hopes were not fulfilled, and he still could not find a permanent, decently paying position, despite his Olympian reputation among geophysicists. There are indications that his Jewish background played a part in all this."

Hertha herself provided a somewhat different view of her husband's decision to emigrate. Her husband did indeed foresee trouble in Germany, but not from the Nazis: "it was too early for that, you see." Instead, Beno Gutenberg feared another world war was imminent. He and Hertha had two children: he didn't want them to grow up in the midst of war. Hertha also described the bleak financial situation in Germany, and not just for underpaid scientists: raging inflation eroded the buying power of the mark, and food remained scarce. "I should tell you how terrible the hunger was," Hertha remarked. The hard currency dissolving before their very eyes, people took to bartering for the necessities of life: soap for a pair of shoes, meat for soap.

The Gutenbergs arrived in California in 1930 with Arthur and Stephanie as well as Beno's mother. Beno's mother made the arduous journey back to Germany, not convinced her move should be permanent. But soon after she returned home, she decided that she didn't care for Germany any more; she wanted to return to be with her son in California. The entire Gutenberg family made the journey back to Germany in 1932 to collect her. By this time, ugly handwriting was on the wall: the Gutenbergs went to Salzburg for the Mozart festival and found it "full of Nazis. We all felt very bad about that. My husband said, "You see now." In Darmstadt they witnessed children no more than eight or nine years old, digging holes where they could spend the night underground.

By the time the Gutenbergs made it as far back west as San Bernardino—perhaps forty miles east of Pasadena—the children shouted, "Here is California!" and counted the palm trees from the train. (Those same palm trees, or perhaps their figurative descendants, sent the exact same message to the author on childhood visits from the East Coast to California. There is something about a soaring palm tree that makes a profound impression on a child who has not grown up with such bizarre and iconic specimens.)

Gutenberg arrived at Caltech just ahead of another, far more famous, Jewish scientist from Germany. Albert Einstein arrived at Caltech as a visitor in 1931 with his second wife, Else, two years before accepting a permanent position at the Institute for Advanced Study at Princeton. A natural bond united Gutenberg and Einstein, both brilliant scientists of German extraction. "At least they were on good terms," Richter said, revealing a lack of direct information that suggests that the two men were more friendly with each other than either was with him. (One cannot appeal to a language barrier to explain the lack of closeness: Richter spoke German as well, having grown up with the language to some extent, although his family had spoken primarily English at home.)

Hertha's recollections reflect clearly the warmth of the relationship, not only between the two scientists but between the two couples as well. The Einsteins were regularly "wined and dined." Hertha explained, "He really liked very plain food. Very often I picked [Else] up on Saturday morning, went shopping, and she bought some lamb chops or something very plain. She cooked it at my house and brought it over to the Athenaeum, and they ate in their room." The Athenaeum, known within Caltech environs simply "the Ath," was and is the Caltech faculty club, where Albert and Else stayed during the visit. Rooms at the Ath are nicely appointed but not equipped with cooking facilities; visitors generally ate first-rate meals in the dining facility on the premises. The Ath menu does not, however, tend to feature entrees such as plainly baked lamb chops.

Richter relayed one anecdote about his colleagues: walking back across campus following a physics seminar, Gutenberg and Einstein were chatting, "mostly about earthquakes." Another professor came up to them and asked, "Well, what do you think of the earthquake?" "What earthquake?" came the reply. When told the answer, Einstein reported laughed that it had been just another thing that had escaped his attention. The quake occurred just before six o'clock on the evening of Friday, March 10, 1933: the magnitude 6.4 Long Beach earthquake had struck thirty-odd miles from Pasadena, leaving serious damage and not insubstantial loss of life in its wake. Richter noted that the temblor was not especially strongly felt in Pasadena, generating a level of shaking that might not be obvious to people walking outside at the time. Still, the shaking was strong enough to cause tree limbs and power lines to sway—the sort of thing a person might notice if he were aware of his surroundings. Richter later heard the story from the professor who encountered Gutenberg and Einstein on campus, but when Gutenberg arrived at the laboratory he relayed the story himself "with considerable amusement." The wives shared a bit of amusement themselves during a shopping trip the following morning: "What do you say about our two dumbbells?" Elsa asked Hertha.

"Dumbbell" applied about as aptly to Beno Gutenberg as it did to Albert Einstein. Photographs reveal Gutenberg to have been of notably short stature—he crested five feet by a good inch, perhaps two—and nearly bald during much of his adult life; they also radiate a palpable sense of intensity and intellect. A colleague would later memorialize Gutenberg as nothing less than "the foremost observational seismologist of the 20th century." If his name had been attached along with Richter's to the magnitude scale he helped develop, it would be a household word today. It wasn't and it isn't—a fact that engenders a certain amount of lingering resentment among some that Gutenberg's public acclaim was not commensurate with his accomplishments. (The corollary is clear if unspoken: Richter's fame in a public arena was also not commensurate with his scientific contributions.)

In scientific circles, Gutenberg remains revered as a founding father of prodigious intellect and productivity. He published hundreds of scholarly papers and seven books. His contributions include a precise determination of the depth of the earth's core and illumination of many other properties of the deep earth. (To not shortchange another founding father of seismological science, one should add that geologist Richard Oldham estimated the depth to the core earlier, in 1906, to be 2,550 kilometers. Gutenberg's estimate improved this value by about 15 percent.) In these studies, Gutenberg helped pioneer many of the techniques that seismologists still employ to study the deep earth. Just as a CAT scan images the human body by looking at the pattern of electromagnetic waves that travel through it, the waves from large earthquakes can be used to image the earth. In particular, as waves encounter boundaries within the earth, for example that between the liquid core and the overlying solid mantle, they can be reflected or refracted (bounced or bent) in ways that can be predicted from wave theories. Employing clever techniques to quantify the pattern of waves recorded at the surface, seismologists can infer a great deal about the nature of key layers, which in turn tells us much of what we know about the structure of the planet beneath our feet. Gutenberg's studies generally focused on what seismologists call *teleseismic* data: waves from large earthquakes that travel through the entire earth and arrive at distant locales with amplitudes far too low to be felt, but large enough to be detectable on certain types of seismometer.

Gutenberg also teamed up with scientists at the Seismo Lab, in particular Richter, to work on matters of both local and global importance. As is well known in seismology circles, but not among the wider public, he made contributions to the initial development and later refinement of the scale that would bear Richter's name alone. That it did not become known as the Richter-Gutenberg scale, or vice versa, is a story that deserves a chapter all its own.

Gutenberg's name did become indelibly intertwined with Richter's—in scientific circles if not the popular lexicon—via another of the pair's seminal contributions, one that involved the overall statistics of earthquake occurrence. The so-called Gutenberg-Richter (sometimes simply G-R) distribution describes the overall predictable pattern, or distribution, of earthquake sizes that occur in a given region. Essentially, in any region on

Earth one finds ten times as many magnitude 4 quakes as magnitude 5's, ten times as many 5's as 6's, and so on. The simple rule breaks down only when one gets up to magnitudes that cannot occur because there are few (or no) faults large enough to generate them.

As it happens, enlightened seismologists nowadays acknowledge that the seminal 1944 publication by Gutenberg and Richter was not quite so seminal after all. As we will see, Richter himself described a basic version of the result in a 1935 paper. A team of Japanese seismologists, Ishimoto and Iida, also published the observational result in 1939. One might wonder if political events during the intervening years had anything to do with the lack of acknowledgement of the Ishimoto and Iida results in Western circles. However this sort of discovery and rediscovery remains common in science, where fields move forward collectively and the next important step not infrequently occurs to more than one person or group at about the same time. (Truth in advertising obliges the biographer to note that Charles Richter did make a point of following the Japanese literature, and abstracts—if not entire papers—were generally published in English.) In any case, while modern Western seismologists generally do now reference Ishimoto and Iida's 1939 publication ahead of Gutenberg and Richter's 1944 paper, the name "G-R distribution" has stuck, at least in Western circles.

To embark on a brief seismological tangent, the distribution of earthquake magnitudes remains very much at the forefront of modern seismological research: Why does the simple pattern hold? *Does* it always hold? How can it be reconciled with the observation that a very large fault like the San Andreas seems to produce only very large earthquakes—for example, a magnitude 8 every one hundred years—but not a commensurate number of 7's, 6's, and so forth? The answer to this last question seems to hinge on the fact that overall earthquake distributions are never calculated for a single fault, but rather for a region that contains an assemblage of faults. Given the enormous complexity of both the earth and earthquake processes, it is mostly surprising that the "G-R distribution" is so simple, and so universal.

The G-R distribution can also come in handy for quick estimations of how often earthquakes of a given magnitude will strike. Worldwide, the planet experiences one magnitude 8 or greater earthquake every two to three years; one therefore expects a magnitude 9 or greater quake every few decades and a magnitude 7 or greater every few months. Southern California, meanwhile, experiences magnitude 6 or greater every two to three years, therefore magnitude 7 or greater every few decades, and magnitude 5 or greater a few times a year. Earthquakes do not, of course, occur like clockwork, but such calculations can successfully forecast the numbers of earthquakes of different magnitudes that will strike a given region, on average. The G-R distribution, modified to account for a number of recognized complexities, therefore remains at the very heart of modern seismic hazard assessment—forecasting the expected level of earthquake shaking, or hazard, in a given region over a given time period. It is striking that the seminal 1944 Gutenberg and Richter publication, "Frequency of Earthquakes in California," occupies all of four printed pages in the *Bulletin of the Seismological Society of America.*

On a personal level, colleagues describe Gutenberg as intensely, passionately, and nearly exclusively devoted to scientific inquiry. Inge Lehmann, essentially the only founding mother of seismology, reported that during her visit with Gutenberg in 1926, she worked alongside him with zero communication of a personal nature. (Lehmann is recognized today principally for her discovery of the earth's inner core: a small, solid iron sphere within a liquid outer shell.) Lehmann described Gutenberg as a wonderful teacher and an unselfish collaborator, and credited him with her excellent introduction to seismology. Shelton Alexander, who went on to a long career at the Pennsylvania State University following his graduate school tenure at the Seismo Lab, describes Gutenberg simply as a "perfect gentleman"—one of a very few he has known in his lifetime.

Notwithstanding a monumental intensity of focus, Gutenberg's colleagues describe him in almost unfailingly glowing terms: noble, kind, gentle, warm; as the Long Beach earthquake vignette illustrates, not without a warm and sometimes wry sense of humor. Former Seismo Lab student John Lett recalled the occasion when Gutenberg's colleagues celebrated his birthday with the ceremonious presentation of a gift: a hairbrush for his nonexistent hair. By Lett's recollection Gutenberg was delighted and proudly kept the brush on display on an office shelf.

In a memorial piece, colleague Robert Sharp wrote of Gutenberg that "his enthusiasm bubbled over and gave generously to all. The twinkle in his eye and the kindly smile were always there." He added, "Not only was he admired, liked, and respected by all, he was literally adored by those who worked closely with him." A few former colleagues do offer recollections that perhaps redeem Beno Gutenberg from sainthood, describing a sometimes autocratic (stereotypically German, some will dare say) approach to his leadership of the lab. When prospective graduate student George Shor balked at Gutenberg's suggestion of a thesis topic, instead suggesting one he considered more interesting, Gutenberg smiled—then switched off his hearing aid and returned his attentions to his own work, leaving Shor to turn elsewhere (to Richter, in fact) to find a thesis advisor. Students also described Gutenberg as a detached lecturer, his concentration focused on filling up blackboards rather than interacting with the class.

A few forgivable foibles notwithstanding, Gutenberg brought enormous charm and intellect to the young Caltech Earth science department and Seismological Laboratory, immediately infusing the institution with energy. Knopoff notes that at the annual meeting of the Seismological Society of America in 1931, six of the fourteen papers presented were from Caltech, three of them authored or coauthored by Gutenberg. Just two years earlier there had been no Caltech contributions among the five presented papers on seismology. With Hugo Benioff and Charles Richter, Gutenberg anchored a seismology program that quickly rose in accomplishment and prestige, in short order becoming one of the world's premier centers for earthquake research.

The move from Germany to California brought Gutenberg from a region where earthquakes were rarely felt to one in which they are part of the fabric of daily life. His research interests continued to reflect a global focus, to the consternation of lab head Harry Wood, who thought their focus should remain on local earthquake problems. Gutenberg did turn his some of his abundant energies to local earthquakes. In these latter endeavors, he perhaps could not have found a better collaborator than young Charles Richter, a talented observational scientist in his own right and one whose professional passions remained largely focused on California earthquakes throughout his own career.

Gutenberg and Richter teamed up to write what Knopoff describes as "four monumental papers" describing the detailed nature of earthquake waves. These publications included a comprehensive catalog of earthquake waves: beyond the familiar P and S waves described in basic seismology textbooks, they include waves whose names reflect not only the wave type but also their route of passage through the earth. A ScS wave, for example, is an S wave that travels down into the earth, is reflected when it hits the core, and travels back up to the surface as a S wave. The nomenclature gets interesting by virtue of the complexity of earthquake waves. For example, S waves cannot travel through the liquid outer core, but an S wave can hit the core, get converted into a P wave, continue through the core as a P wave before reemerging out the other side and being converted back into a S wave. At the surface, such a wave is identified as SKS. Waves can moreover experience multiple reflections at different layers in the earth, leading to ungainly names like ScSSKP. It thus required a momentous effort by Gutenberg and Richter to catalog all such waves.

The pair also, as already discussed, made seminal contributions to local earthquake studies, chiefly the magnitude scale and the distribution of earthquakes. It was a phenomenally successful partnership if measured by the products it produced. It was also, most assuredly, a complex relationship. In a 1983 interview Frank Press described it as a "love-hate relationship": "They didn't like each other but they respected each other. There was competition. These were two strong personalities, each trying to do good work and receive recognition for it. And they were in overlapping fields—for example, the magnitude scale. So it was an uneasy truce, but they let each other live. Once in a while there was a flare-up."

Press goes on to acknowledge that most of Richter's publications were coauthored with Gutenberg. "It was a marriage of convenience," he said. "Everybody recognized it as that." The marriage was not, however, marred by continual strife: "They were two giants who occasionally saw things differently—in the style of operation of the place or how a piece of data might be interpreted. And then they came from entirely different cultural backgrounds. In many places where you have these kinds of people—very strong, very confident people—the atmosphere is frenetic and tense and unpleasant, but that wasn't the case." Nor, Press acknowledges, was the atmosphere entirely pleasant: "There was a rivalry, a lack of friendliness."

This was, one notes, how the relationship appeared to one close observer, a man who not only worked in close contact with both individuals for several years, but also was one of the most talented seismologists of his generation. As with so much else in Richter's life, however, appearances could be deceiving. Judging from Richter's papers, at first glance his relationship with Gutenberg does indeed appear rather formal. Among Richter's papers at the Caltech archives is a folder of correspondence with Gutenberg. Most of the material therein involves Gutenberg's death in 1960: letters that Richter sent and received as well as the memorial piece he wrote together for the Geological Society of America. One finds little personal correspondence between the two men—not surprisingly, as close colleagues would not often have occasion to send letters. (Now that scientists have discovered email, they send each other "letters" all the time—but one wonders if the biographers of the future will ever be able to find any of them.) The folder does contain a small handful of letters that Gutenberg wrote when he was traveling: all almost exclusively focused on scientific matters. Letters as late as 1948 begin, "Dear Dr. Richter," although by 1955 the letters did begin, "Dear Charles," and were signed simply "Beno."

In one two-page letter, dated October 31, 1955, Richter, addressing his colleague as "Beno," discusses differences of opinion regarding issues related to the magnitude scale. The letter opens, "What follows has been written with much care and some trepidation; but I hope it will provide a basis for clearing out most of our remaining differences about the magnitude scales. Please believe that I want to agree with you, and that I have not maintained a divergent opinion out of obstinacy or inertia. I only follow the precept I have learned from you—that one owes it to one's colleagues as well as to oneself not to abandon a position until one is wholly convinced; then one should be able to take the new position wholeheartedly." Richter continued, "Naturally I often take up the other side of a question in debate, simply for the benefit of discussion; but to hold a position when I know you have a different conviction is hard for me, believe it or not; it goes against both my respect for you and my general dislike of conflict."

The letter then launches into the meat of the discussion, which to a large extent concerns technical matters related to refinements of earlier magnitude scales. In particular, Gutenberg and Richter had been working to develop a "unified magnitude scale" that could incorporate the original "Richter scale" as well as later developments, and be used to determine magnitudes of earthquakes large and small, near and far. (Most seismologists today are inclined to view these developments as Gutenberg's rather than Richter's contribution: by Shelton Alexander's account, however, Richter was a "full partner" in later efforts.) To a large extent the debate swirled around nomenclature: what this new scale should be called, given other definitions and terminology already in use. Notably, Richter wrote, "Unified magnitude should be its official title, used at least in our joint publications; elsewhere I intend to refer to it as "Gutenberg's unified magnitude scale," or for short "the Gutenberg scale." He notes later that "Richter scale," should only properly be applied to shocks smaller than magnitude "5 or perhaps 6." Neither name, "unified" or "Gutenberg," ever stuck. Moreover, the letter in the archives is not a mimeograph copy, thus proving only that the letter was written, not that it was ever sent. The letter does nonetheless provide a measure of illumination regarding Richter's own feelings on the subject of what the world knew as *his* scale. He did feel a sense of ownership of the original contribution—the paper that presented the magnitude scale was indeed his paper—but felt strongly that credit should be given where credit was due in the further development of the concepts. Betty Shor recalls Richter as having been "quite self-conscious" about the notoriety of the "Richter scale." (I shall return to discuss this issue at a later chapter.)

Frank Press—later tapped by Jimmy Carter for the august role of President's Science Advisor—had been in a position to observe the interactions between Charlie Richter and Beno Gutenberg. One is not inclined to doubt his powers of observation, which is to say, he knew how the relationship appeared to the outside observer. Hertha Gutenberg had her own view of the relationship between her husband and his colleague. Hertha plays the Great Lady role, and plays it well, in her 1980 interview with Mary Terall, yet those who knew her in person observe that she had a sometimes sharp tongue. The barbed edge creeps into Hertha's voice during the interview when she described her late husband's collaborations with Richter. "Afterwards," she said, "after my husband died, Richter published one book, a textbook on seismology, but I don't think he did very much research." According to a longtime colleague, she expressed her views more bluntly in private conversations after Beno's death, including her feeling that her husband had not received proper credit for his contributions to the magnitude scale. She also felt that her husband had been overly patient with his rather childish colleague.

Hertha went on to suggest that Richter had little interest in an update of a more scholarly publication, a book that he and Gutenberg first published in 1941. In spite of failing health Gutenberg had set out to revise the book on his own, but never got a chance to finish the job. Having taken ill with severe influenza, Gutenberg was seen by a doctor on Sunday, January 24, 1960, who told Gutenberg he should not be working. At six o'clock the next morning he was taken to Huntington Hospital, where his lungs were found to be full of fluid. Hertha stayed with him throughout the day; he kissed her goodbye when she left. He died at eleven-thirty that night. A phone call relayed the news to Hertha, who understandably "just went to pieces."

Richter was in Japan when Gutenberg died and was unable to return. However, he exchanged cordial letters with Hertha in the month following Gutenberg's death. He extended his sympathies and expressed a sense of "irreparable loss" on both a personal and scientific level. She thanked him for his kindness. Yet her later words revealed unmistakably hard feelings. From Hertha's perspective, Richter was odd and rather arrogant, an immature man who had managed to ride on her husband's substantial coattails. And perhaps most grating of all: as far as the public was concerned, Charles Richter—not her beloved and enormously accomplished late husband—had become the most famous seismologist of all time.

Frank Press looked at Richter and saw a man who had been intensely competitive with his tremendously successful and productive colleague. Another colleague, Clarence Allen, also recalled heated arguments over the locations of earthquakes. "And then almost invariably," Allen said, "an hour later, they would come back and apologize to each other formally." Richter's own words, meanwhile, reveal a man who was not engaged in a competition as much as a furious struggle to keep his own head above water. They reveal how profoundly outward appearances could be at odds with inward realities. Focusing on what Richter himself said on the subject of his colleague, one finds nothing but the deepest respect and admiration. His own words also revealed his continuing emotional struggles that, although perhaps invisible to people like Frank Press and Hertha Gutenberg, very much influenced Richter's actions throughout his life. In the 1949 letter to his doctor Richter described the sense of malaise that he found himself battling—an acute tendency towards procrastination that interfered with his professional responsibilities.

Richter described his position at the laboratory thus: "The . . . routine, which involves a great deal of measurement, filing, and tabulation, is either my lifeline or my chief handicap, I hardly know which. I have a more or less running fight against the pressure of correspondence, research work, students, visitors, and the disposition of others to introduce time-wasting changes and additions into the routine. Here, after twenty years, I have developed an almost belligerent insistence on no postponement and no procrastination; it may be this that tends to make me dilatory elsewhere? If so I don't see why."

Richter went on to describe a dozen research papers that he had started but never completed, the work postponed in favor of other efforts. "The many papers I have turned out," he wrote, "have mostly been in collaboration with Dr. Gutenberg, who is an indefatigable worker himself, and almost a slave driver when there is something he wants to get finished." He continued, "Papers I have worked out myself have usually been very short; material which would be assembled in a few weeks of steady work, written up very briefly with the maximum of data and the minimum of interpretive comment, and sent off promptly to be published."

Just as Frank Reed had done decades earlier, Beno Gutenberg must surely have stoked Richter's anxieties and insecurities. Whereas Reed had been a big man, Gutenberg was notably small of stature, but a towering figure nonetheless. Knopoff described him as "very personable, and lively. He was well organized and kept to a precise daily schedule. Although his scientific demands on himself were rigorous, Gutenberg was (for the most part) gentle and self-effacing in his relationship with others. He was helpful to anyone who asked a question of him and was tolerant of critics. Gutenberg was a man who could give colleagues and students a liberal education in scientific method, made pleasantly easy by kindness, patience, amazing industry, and a delightful sense of humor. He was a cultured individual, well read, and with wide interests."

The contrast with Charlie Richter begins to emerge as very nearly painful: Richter was awkward where Gutenberg was personable and lively, utterly disorganized where Gutenberg had been relentlessly organized. Moreover, while Richter certainly had wide interests, he was not nearly as outwardly worldly as his longtime collaborator. Clarence Allen summed it thus: "[Gutenberg] and Charles Richter were just utter contrast, because Richter was a very different kind of person."

Nor did Richter share his colleague's musical talents: Gutenberg played the piano and organ, and sang in a synagogue choir in his early years in Germany. Gutenberg's interests in music were part of his overall kinship with Albert Einstein, who played violin in chamber music events at Gutenberg's home during Einstein's years in Pasadena.

The painful contrast extended to their families as well. Among the adjectives used to describe Hertha, nowhere does one find, "peculiar"—the word she herself would use to describe Lillian Richter. (Appearances could be deceiving, however, with Lillian as well as Charles Richter, a story we return to in a later chapter. For now, however, one might recall an earlier observation: Hertha's tongue could be sharp.) Hertha and Beno remained a devoted couple until his death in 1960. A photograph of her taken in 1980 reveals a woman whose beauty and forthright intellect were scarcely muted by her eighty-three years. Whereas the Richters had no children together, the Gutenbergs had two, by all accounts well adjusted and successful in their own right. After serving in World War II (on the American side), Arthur received his Ph.D. from Stanford; Stephanie went to Berkeley and, later, to Stanford Medical School.

Whereas the Gutenbergs and Einstein clearly established a close friendship, Richter—the man who had aspired to a career in modern physics—later acknowledged that he had had very limited interactions with Einstein during the latter man's years at Caltech. "I always felt I wasn't taking advantage of the opportunity I might have had," he said. He then described the routine: The Caltech calendar would regularly include an item, "Physics seminar, subject and speaker to be announced." This speaker, Richter went on to explain, "was nearly always Einstein, that was understood. They preferred not to put it definitely on the calendar because it tended to attract cranks and curiosity seekers." But, even though Richter was aware of these things, "I failed to attend most of those. I was pretty well absorbed in my time at the Laboratory and other things going on outside." One cannot help but wonder, however, if he didn't shy away for more personal, and more painful, reasons.

In the end, the long relationship and partnership between Richter and Gutenberg mirrored the complexity of the men—of Richter in particular. As for his feelings towards his colleague, the definitive authority, Richter himself, described it himself in his 1979 interview with Ann Scheid: "I owe a very great deal to him and came to regard him with almost a filial affection. I was very, very fond of that man." Those who observed Richter's interactions with his colleague saw only the outside of the story; as was so often the case, they never had the opportunity to hear Richter's inner thoughts. For Richter the relationship was not a marriage of any sort but rather something else entirely. In the diminutive form of Beno Gutenberg Richter found the one real father figure of his life: not the strapping, big man Richter had yearned to be as a boy, but in every other respect an embodiment of his own, long unfulfilled, idealization of manhood. Brilliant, accomplished, personable, comfortable in his own skin, happy with his wife and children—Gutenberg possessed all of the right talents to rise to preeminence in his chosen field of study; not only a shining intellect but also curiosity, passion, drive, and formidable organizational skills. Few seismologists today would fail to credit Gutenberg as the more broadly accomplished, and preeminent, scientist of the pair.

Gutenberg's drive and organizational skills provided the impetus that Richter needed to harness his own horses, as they say—and set his own demons far enough aside to make his own tremendous contributions in science. Yet at the same time Richter brought his own assets to the team: profound scientific insight, tireless energy and tenacity in routine analysis and interpretation, and what longtime colleague Bob Sharp called the ability to see where and how complicated data "fitted into some larger picture." Observational seismology still requires a unique collection of talents: the ability to digest hopelessly complicated data and extract useful information from a tangled heap of wiggly lines. In Richter's day this involved a staggering degree of tedium as well: the analysis of large volumes of data without calculators or computers. Richter's natural intellectual gifts were such that he did not contribute to the design of early seismometers, nor was he as effective as Beno Gutenberg at driving the field forward at a relentlessly steady pace. He did, however, possess a most unique set of talents and proclivities that allowed him to become the key bridge between the two:

the true observationalist, the link between the mechanical world of instruments and the scientific world of cutting-edge conceptual science. He also had tremendous scientific insight as well as a far stronger grasp of theoretical seismology than most of his colleagues realized, as his students were sometimes surprised to learn when they signed up for classes that they expected to focus on purely observational aspects of the discipline.

Richter also brought to his collaboration with Gutenberg his deeply rooted passion for local earthquake problems. Prior to his arrival in Pasadena, Gutenberg's research had focused almost exclusively the structure of the deep earth. Once he arrived in Pasadena, spurred by both the abundance of high-quality recordings of local earthquakes and Richter's abiding interest, he turned his attentions to problems at a different scale. Local earthquakes can be a parochial matter, not of profound scientific importance. Yet sometimes, "local earthquake problems" can involve fundamental questions regarding earthquake processes. (A degree of bias nonetheless exists in seismology circles. Some have quipped that "deep earth studies," which is to say research focused on the deep structure of the planet, are considered somehow more deep, which is to say profound, than studies focused on the shallow earth, including local earthquake studies.) Some of Gutenberg's work with Richter, for example the identification of the Gutenberg-Richter distribution of magnitudes, was—and remains many decades later—of seminal importance for earthquake science as well as hazard assessment.

One measure of the durability of a scientific contribution is the so-called citation index: the number of times that a paper is cited in subsequent scholarly works. Of the papers on which he was lead or sole author, Gutenberg's most highly cited contribution is his 1944 article with Richter. Although Gutenberg authored far more papers than Richter, few come close to the number of citations for this one jointly authored article. None of Gutenberg's papers are as highly cited as Richter's 1935 article describing the original formulation of the magnitude scale. What's more, none of the papers written by either man or by both collaboratively come close to the number of citations garnered by Richter's landmark textbook, *Elementary Seismology.*

Gutenberg and Richter teamed up to write another classic book, *Seismicity of the Earth*, that was published in 1949 and went on to be extremely well cited. This book remains even today a valued reference for seismologists who investigate earthquakes on a global scale. Although much has been learned since 1949, remarkably little in this book has been proven wrong, and in some cases it has proven more right than conventional wisdom in the field many decades later. For example, in 1978 a pair of Seismo Lab graduate students, Seth Stein and Emile Okal, became interested in the so-called Ninetyeast Ridge, a long mountain chain in the Indian Ocean. At the time, the geologists viewed this feature as an "aseismic ridge": an underwater mountain belt that does not produce earthquakes. Stein and Okal realized the ridge only appeared to be an earthquake-free zone (they themselves would not have used this term) because maps at the time showed only those earthquakes that had struck since 1963. They then discovered, to their surprise, that Gutenberg and Richter had discussed a band of large earthquakes along the ridge in *Seismicity of the Earth*, even though the undersea ridge itself remained undiscovered in 1949. In effect, Gutenberg and Richter's keen and exhaustive analysis of limited early data had found the ridge before it was found. Graduate student Douglas Wiens later worked with Stein and Okal, using results from a textbook of a vintage that would mark most science texts as archaic, to finally make sense of this enigmatic feature. "Gutenberg and Richter were incredibly perceptive," Stein observes. "Their book, written long before plate tectonics was discovered, helps us solve tough tectonic problems by giving us a longer earthquake history." Gutenberg, not Richter, is the lead author of *Seismicity of the Earth*, yet Stein is inclined to suspect that Richter's handiwork is very much evident in the encyclopedic nature of the book—the very reason it remains so valued today.

Gutenberg's star might shine more brightly in seismology circles, but there can be little doubt that Richter's star shines very brightly in its own right. Richter contributed his fair share of talents, both inspiration and perspiration, to the partnership. In the fireball form of Beno Gutenberg, Richter found not only his match but also the motivation that he could not always summon on his own. But as his own words make clear, he found something else as well. Hertha Gutenberg may never have known how close she came to the mark in regarding her husband's quirky young protégé as childish.

CHAPTER 8

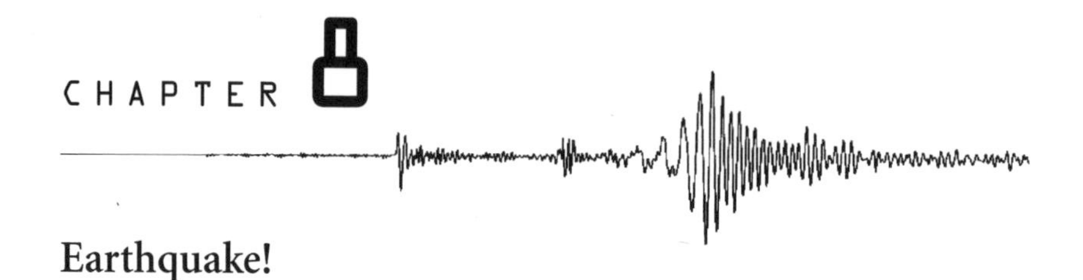

Earthquake!

> This calamity had a number of good consequences. It put an end to efforts by incompletely informed or otherwise misguided interests to deny or hush up the existence of serious earthquake risk in the Los Angeles metropolitan area.
> —*Charles Richter,* Elementary Seismology *(1958), p. 498.*

THE FIRST earthquake that Charles Richter felt made an impression on him as a boy, but only to a point. It would have made for a good story, had the magnitude 5.5 temblor of May 15, 1910, inspired a lifelong interest in seismology, but it did not. As we have already seen, seismology was for Richter far more acquired taste than childhood passion. Having stumbled unwittingly into the field of seismology at the age of twenty-seven, Richter arrived on the scene two years after one of the important earthquakes in the early history of Southern California, a magnitude 6.3 temblor that struck at 6:44 in the morning of June 29, 1925, near Santa Barbara. Wood-frame homes in the area rode out the shaking relatively well, although many chimneys were lost. The quake took a far heavier toll on commercial buildings in the city's downtown area, damaging or destroying many masonry structures. Shaking was felt (weakly) as far as Watsonville to the north and San Bernardino to the east.

Whereas business leaders had endeavored to paint the 1906 disaster in San Francisco as chiefly one of fire rather than earthquake, the same could not be done in 1925. The Santa Barbara earthquake produced no fire: the $8 million in property damage and thirteen deaths could only be attributed to the destructive wrath of the earthquake itself. Insurance companies found themselves facing heavy obligations. Many companies immediately raised rates on future policies; others stopped offering earthquake insurance altogether.

Richter most likely had only passing awareness of any of this at the time, beyond what he may have read in the newspaper or heard in the corridors of Caltech. He would become well versed in the story, however, after joining the Seismo Lab. His papers in the Caltech archives include a thick folder on the earthquake, including many photographs of damage. The Santa Barbara earthquake was an important early chapter in the development of building codes in California, a prime concern of Richter's later career and life. In particular, damage from the temblor highlighted in dramatic fashion the deficiencies in the building practices of the time: the reliance on masonry construction, even for bearing walls, with no reinforcement or lateral bracing to help walls withstand shaking. The earthquake also exposed a less obvious hazard: the potential for abutting but separate buildings to damage each other as each responded independently to the shaking of the ground.

The 1925 earthquake may have found its way into Richter's awareness only after the fact, but by the time the Long Beach earthquake struck at 5:54 p.m. on March 10, 1933, earthquakes were very much on his radar screen. He was living with his wife in Los Angeles at the time the earthquake struck but had planned to attend a chess club meeting in Pasadena that evening, and so had stayed late at the laboratory. His colleague, Hugo Benioff, and an assistant were in the lab as well. Most of the seismograph recordings were still done on film, requiring a laborious process of development before one could inspect a seismogram. Unable to field queries from journalists after earthquakes were felt in the area, Richter and his colleagues had fashioned a ink-pen seismograph that could be inspected immediately. "It was a relatively crude system," Richter wrote, "but adequate for emergencies."

On the unusually warm evening of March 10, 1933, the earth presented the lab with one heck of an emergency. Richter noticed the deflection of the seismograph pen just as he felt waves from the earthquake shake the building. Anyone who has felt a number of earthquakes, whether he or she is a seismologist or not, soon gains an appreciation for the different nature of shaking from earthquakes of various magnitudes felt at different distances. A nearby magnitude 3 temblor might produce a rude but brief jolt, while waves from a distance magnitude 6 earthquake can feel powerful and alarming even when they are in fact diminished by distance to nearly harmless levels. The seismologist can draw on a mathematical and physical understanding to explain such differences. For example, just as large musical instruments produce sound waves with low tones, so too do larger earthquakes generate powerful waves with very long periods ("tones"). The details are unimportant here, but from the nature of the very first waves to reach Pasadena from Long Beach, Richter would have had no doubt that they had been generated by a strong earthquake—one centered not too far away.

"With appropriate remarks," he later wrote (suggestively, but without further elaboration), "I headed for the recorder—which shortly ran out of ink and was restored to operation only with some trouble." Attempting to decipher the seismogram—what they had of it—Richter and Benioff found the recording too messy to provide many clues about the event they had just experienced. Shaking from such a powerful nearby earthquake would have blasted their recording devices off-scale, rendering them all but useless to determine the distance to the epicenter. At the time, the only practical earthquake-location procedure required determination of the time that both the P and the S wave arrived at a given station. From this separation one can work out distance, just as the delay between lightning and thunder reflects the distance to the lightning bolt. When early instruments went off-scale, the later S waves could not be distinguished. The seismograph did, however, continue to record aftershocks, which are known to cluster in proximity to the mainshock. From these smaller shocks Richter and his colleague were able to distinguish separate P and S waves, which allowed them to estimate the distance of the earthquake from Pasadena.

When an earthquake occurs in Southern California today, data from a network of seismometers are beamed via microwave links, phone lines, and Internet connections to the Seismo Lab, where computers spit out initial magnitude and location results within minutes. Earthquake locations can now be determined using sophisticated algorithms from P-wave times alone. However, even as late as the mid-1960s, analysis was still done largely by hand, by scientists and graduate students. When a moderate earthquake rattled a good part of Southern California in 1970, Thomas Hanks and other younger scientists at the lab found themselves momentarily flummoxed by saturated seismograms. Richter, who had been within earshot although not part of the conversation, was heard to grumble, "Use the S-minus-P times of the aftershocks"—the same technique that he and Benioff had turned to nearly four decades earlier. It is said that necessity is the mother of invention, and without question the pioneering observational seismologists of Richter's day were inventive—they had to be.

Richter's sage wisdom lived on in ways he probably never would have anticipated. In 1975 Hanks used Richter's method with early seismograms to obtain a reliable estimate of the location of a moderately large earthquake that had struck near Lompoc, California, in 1927. Hanks's location differed from that of another scientist who had analyzed seismograms from the earthquake recorded on seismometers around the globe. The "teleseismic" solution placed the earthquake on the Hosgri Fault, a noteworthy result because nearby Diablo Canyon was under consideration as a site for the nuclear power plant that would be built in 1985. Locations from global data are notoriously imprecise, however: the local data, combined with Hanks's recollection of Richter's grumbled advice, won out.

In 1933, basic data analysis posed even more of a challenge than in 1970. In 1933 seismometers were stand-alone devices, unable to communicate with anything but their own internal recording media, usually film. Estimating an earthquake location, meanwhile, requires observations from seismograms recorded in at least three different locations. With a single instrument one can estimate distance, but it would be some time before the data could be analyzed to provide a precise estimate of location.

In the meantime seismologists often drew on other evidence to pinpoint the location of a large earthquake. Some large earthquakes extend all the way to the surface of the earth, creating a telltale scar that leaves no doubt where the quake occurred. The 1933 temblor left no such signature but did leave an unmistakable pattern of death and destruction, which with some exceptions were concentrated near the seat of the disturbance. In fact, with only a single seismometer at their immediate disposal during its early days—and thus no way to pinpoint a precise location—the "experts" at the Seismo Lab would ask callers about the extent of damage in their areas and thereby get a sense of a temblor's location.

Echoing Santa Barbara's experiences eight years earlier, many unreinforced masonry buildings failed catastrophically on that March evening, including many public school buildings in and around the city of Long Beach. Parents could not look at the destruction without feeling their hearts in their throats at the thought that the quake might have just as easily struck three or four hours earlier. Tragically, parents in northern Pakistan did not have to imagine the horror when the magnitude 7.6 Kashmir earthquake of October 8, 2005, struck just before nine in the morning local time, after the school day had begun: thousands of children were killed when many poorly built school buildings collapsed. The children of Southern California were to a large extent spared in 1933, but 120 people died in the earthquake, including 5 children killed in gymnasium buildings. The temblor caused an estimated $50 million in property damage—in depression-era dollars.

At the Seismo Lab that night, Richter's hopes of a chess club meeting were dashed. (He would arrive home very late that night to Lillian's report of the cat's eloquent response to the temblor: the animal had "spat on the floor because it wasn't behaving properly.") Harry Wood arrived soon after

Fig. 8.1. Damage to a public school building caused by the Long Beach, California, earthquake in 1933. (Photo courtesy of National Geophysical Data Center.)

the earthquake with a ticket that would go unused, for that evening's performance of a play, *What Happened to Jane.* Gutenberg arrived soon as well, relating the story that would become infamous—how he and Einstein had managed to not notice the shaking as they strolled across the Caltech campus together. Gutenberg did, however, feel some of the aftershocks, to his elation: "opportunities" to feel earthquakes having been few and far between in his native Germany.

Soon after the earthquake Richter embarked on investigations in the field, recording aftershocks on a portable seismometer that he set up in different areas to measure the degree of shaking. Such studies remain useful today, for example to understand how the pattern of shaking during a big earthquake is controlled by the local geology at any given site. Portable seismometers nowadays are far more sophisticated than their forerunners in Richter's day, but any "aftershock chase" remains, even today, a time-consuming endeavor, like the one that took Richter out of the laboratory for many hours in 1933.

Harry Wood remained in the lab to deal with officials, the media, and the public. Anxiety remains high in the aftermath of any damaging earthquake. "As always after a strong earthquake," Richter wrote, "many persons, including officials, were almost pitifully eager to be assured that the

Fig. 8.2. Damage to a public school building caused by the Long Beach, California, earthquake in 1933. (Photo courtesy of National Geophysical Data Center.)

shaking would soon cease. Such assurance could not be given; as was to be expected, occasional perceptible aftershocks went on for months."

As earthquakes go, aftershocks are remarkably well behaved. Their patterns of occurrence follow simple empirical rules that were, for the most part, well established even in Richter's day. On average, for example, the largest aftershock of any given magnitude 6 earthquake will be of magnitude 5. The aftershock sequence will also include ten events of magnitude 4, one hundred of magnitude 3, and so forth—the familiar Gutenberg-Richter relation that describes the distribution of magnitudes of earthquakes in a given area over time. Of course these numbers represent a simplification—for example one gets not only 3s and 4s but also 3.1s, 3.7s, and so on—but they convey the general idea. The rate of earthquakes, moreover, follows a simple curve on average, diminishing with time. The rate after ten days will be about one-tenth what it was after one day; after one hundred days it will be one one-hundredth of the rate on the first day.

One common misperception about aftershocks persists to this day, to the unending consternation of seismologists: the overall rate of aftershocks diminishes with time but the distribution of aftershock magnitudes does not change. That is, even late in a sequence, one still expects one magnitude 4 for every ten magnitude 3's. Thus, larger earthquakes become more infrequent as time goes by, but an aftershock sequence can continue for months or even years after people generally stop feeling them. As long as the sequence continues, large aftershocks remain possible. Indeed, a strong aftershock on October 2, 1933, caused further damage, especially to buildings that had been damaged by the mainshock and had not yet been fully repaired.

At the same time, Richter described a "regrettable tendency of the news media, to some extent in print but conspicuously over the air, to suggest that 'another big one' was expected momently." Harry Wood unwittingly fueled this sense of panic with remarks at a conference about the possibility of another large earthquake in the future. Although he had been referring to the entire Southern California region, some in the media took his comments to mean Long Beach and Los Angeles specifically. In fact, the greater Los Angeles area would remain remarkably quiet for nearly four decades following 1933, with no earthquakes even as large as magnitude 5 occurring until the quiet was broken (rudely) by the magnitude 6.7 San Fernando earthquake in 1971.

Fears abounded in the aftermath of the 1933 quake, including those generated by a local radio station's warning of a huge sea wave approaching the coast. In the aftermath of the devastating 2004 Sumatra earthquake, the world is all too familiar with the potentially destructive phenomenon known as a tsunami. These waves are generated by certain types of offshore earthquakes and, sometimes, underwater landslides, but not by an earthquake such as the 1933 temblor. The 1933 earthquake occurred on the Newport-Inglewood Fault, a strike-slip, or lateral, fault located mainly on land. Tsunamis are generated when a large volume of water is suddenly displaced vertically, sometimes by a landslide but more commonly following an offshore subduction zone earthquake that involves a suddenly lurching and uplift or down-dropping of the seafloor. (Southern California is exposed to myriad natural disasters: tsunamis are one of them, but most experts would not rank them high on the list of dangers.)

In 1933 as today, a destructive earthquake in Southern California generates aftershocks well beyond those that register on local seismographs. As Richter described in *Elementary Seismology*, the Long Beach temblor hammered home the message of the 1925 quake, and then some. In classic Richterese, he wrote, "Desperate attempts . . . [purported] to show that there was no serious risk of a destructive earthquake in Southern California—or at least in Los Angeles. The disastrous Long Beach earthquake of 1933 relieved seismologists of the need to argue the matter." The 1933 quake also provided dramatic further illustration of the hazard posed by older, unreinforced buildings. Newer, well-engineered structures performed relatively well; older buildings did not.

The staggering devastation to school buildings left an especially indelible impression. Some raised the question of construction standards: a number of relatively new and costly school buildings had apparently sustained worse damage than other modern structures. Risks of various sorts are inescapable, and humans accept certain risks as part of the bargain of everyday life—even nontrivial risks such as the danger of being killed in an automobile accident. (Americans have, somewhat astoundingly, about a one in one hundred chance of meeting their maker in a car crash.) The chance that one's child might be killed, unnecessarily, at school is not a risk that people accept. California assemblyman Charles Field led the charge to enact a law known as the Field Act. The law was passed on April 10, 1933–just one month after the earthquake. Under the Field Act, the state of California was empowered to review and approve all public school plans prior to construction, with very strict subsequent supervision during construction. The Field Act remains today one of the most stringent earthquake construction provisions anywhere in the world.

The Field Act works. To date, no public school built in California since April 1933 has failed in an earthquake. Some Californians—seismologists and the public alike—do have reservations about the buildings that the Field Act does not cover, including private schools, public school buildings built prior to 1933, and public universities. One ironic demonstration of Field Act inconsistencies played out in the city of Berkeley towards the end of the twentieth century, when a public school for deaf children was closed because the building was deemed unsafe under the Field Act, but the building could then be used without reinforcement to house university students.

The Field Act stipulates that if older school buildings are inspected and deemed unsafe, the consequences of continuing to use them becomes the personal responsibility of local authorities. As Richter described, this provision did not achieve its goal: when schools were found unsafe, communities called bond elections to raise the necessary funds. When the bond propositions were voted down, local school boards were absolved of future responsibility, and the buildings remained in use. This game lasted until 1965, when the state attorney general ruled that a failed bond proposition did not absolve local officials of responsibility. Ironically, many buildings were then reinforced, or retrofitted, at a far greater cost than would have been incurred had the work been done earlier.

The Field Act may be imperfect, but California's children are about as safe in their public school buildings as anyone can be in any building. A law as stringent and costly as the Field Act would almost surely never have been passed had the 1933 earthquake not seared stark imagery of death and destruction—and the mental imagery of horrors narrowly averted—into the collective public consciousness. The temblor had a profound affect on seismologists as well, Richter chief among them: it convinced them beyond all shadow of doubt of the immediacy of the earthquake problem in Southern California, and of the hazard posed by especially vulnerable buildings.

Ironically, as noted, the Long Beach earthquake proved to be anything but a harbinger of worse events to come within the greater Los Angeles region. No earthquake of any consequence struck again during the nearly four decades that witnessed the explosive growth of Los Angeles and its environs. For many residents of Southern California, especially the large fraction transplanted from other regions, earthquakes began to slip quietly out of collective awareness.

However, while earthquakes did not strike the Los Angeles region in the decades following 1933, they certainly occurred elsewhere in Southern California. In 1940 a large temblor struck the Imperial Valley, causing damage in communities like El Centro and Brawley, as well as Mexicali in Mexico. This temblor did leave a dramatic surface scar, one that tore straight across the international border.

Earthquakes such as the 1940 event, as well as other moderate shocks in the sparsely populated desert regions of Southern California, occupied Richter's attentions from time to time throughout his career. Until 1971, however, none of these events was of tremendous consequence for the Southern California region. The 1933 Long Beach earthquake stands as the one earthquake that caught the attention of both Charles Richter and the citizens of Southern California. Nobody in the greater Los Angeles region that March evening could escape the point: like the rest of the state, Southern California is earthquake country.

For the thirty-two-year-old Charles Richter the lessons were even more profound. This one earthquake imparted lifelong lessons regarding the nature of the so-called earthquake cycle, the perils of poorly considered remarks to the media, and the critical need to address earthquake hazard by focusing on the buildings most likely to fall down when a large temblor strikes. The first two topics are reserved for a later chapter devoted to earthquake prediction. The last issue, that of vulnerable buildings, has already been addressed here; the issue of hazard mitigation became a lifelong crusade for Charles Richter. He did not work directly on the building codes that ensure the safety of new buildings. Rather, he was a most effective spokesperson for earthquake hazard mitigation, a voice that continued to communicate the issues to the public and public officials.

When the Long Beach earthquake struck, Richter was a young seismologist who had not ever felt any special calling to the field in which he found himself employed, either with respect to pure science issues or societal issues. After the Long Beach temblor rocked the Southland, leaving communities and lives in shambles, it cemented Charles Richter's attachment to the field that he had never really chosen, but that had chosen him—and that he came to care about deeply . . . perhaps as deeply as anyone.

CHAPTER 9

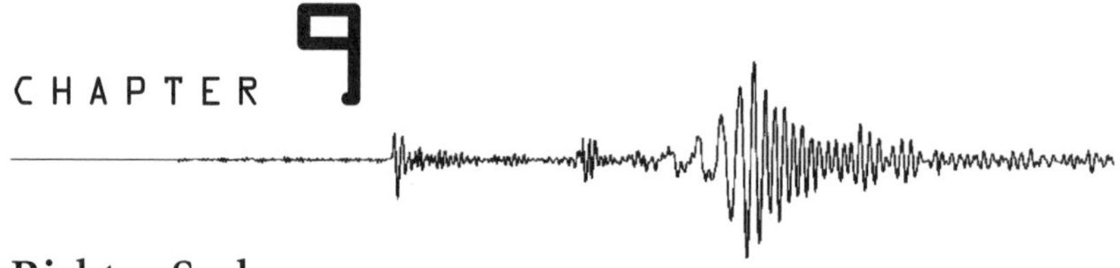

Richter Scale

> I cherish the hope that, if I ever arrive at a fairly satisfactory personal solution, I may be able to make a contribution of permanent value.
> —*Charles Richter, November 13, 1926*

EARLIER CHAPTERS begin to fill in the enormous gaps found in virtually all brief biographies of Charles Richter: "Richter was born on a farm in Hamilton, Ohio in 1900. In 1935 he invented the Richter scale." Now, at last, we move forward to discuss his most enduring scientific contribution; the reason why most people care about him enough to have picked up this book in the first place. His myriad other scientific contributions aside, Richter's stature in the public arena is clearly due entirely to his scale: The Richter scale. A household word even though many still misunderstand the very nature of the scale; even though it was in fact extended—some would say supplanted—years ago within the seismology community.

Biographies say the scale was invented in 1935. In fact Richter used his scale to size up earthquakes in Southern California as early as 1932; the 1933 Long Beach earthquake was among the earliest large events that he studied. The scale was, moreover, not "invented" in 1935; rather the paper describing it was published in that year: "An instrumental earthquake magnitude scale," which occupied the first thirty-one pages of the January 1935 issue of the *Bulletin of the Seismological Society.* The very word "invented" conveys the impression of a mechanical device, which as we will see shortly, this scale is not.

Considering the turbulence of Richter's early adulthood and his circuitous path into the field of earthquake science, one is able to understand the timing of his most enduring scientific contribution. Through his early to mid-twenties Richter struggled with his demons of various shapes and sizes; only towards the end of his third decade did he begin to come to grips with them. By his own account he never reconciled his deep need for artistic expression with his scientific inclinations and aptitudes, but rather reached an uneasy detente. The essence of this rapprochement was, not balance, but very nearly the opposite. By turning the full brunt of his boundless energies to scientific matters, he at last found some sense of equanimity. By his thirties Richter was married, employed, and beginning to gain a measure of professional respect among his colleagues—not in his chosen field, perhaps, but in a field that had chosen him—even as, to some extent, he remained a walking assemblage of contradiction and personal conflict.

Reading the story of the Richter scale, one must keep in mind that computers as we know them today were decades away from invention in the early 1930s. The word "computer" was part of the lexicon, but in carbon-based rather than silicon-based form: a "computer" was a person who "computed." Basic arithmetic calculations could be enormously tedious; the production of a simple graph required laborious efforts by hand. Richter relied on analysis of seismograms of earthquakes in Southern California to develop his scale. Today the equivalent analysis could be done by a competent graduate student in an afternoon, maybe two. In the 1930s it required, in addition to the inspiration, the perspiration of a sustained and focused effort. It required, in effect, the kind of relentless energy and tenacity that Richter possessed, and provided an effective outlet for these traits—traits that, left unbridled, were clearly wont to wreak havoc on his psyche.

But before one can talk sensibly about the Richter scale, one must first step back to address a few more general issues and questions. For starters, how does one size up an earthquake? To answer this one must first consider an even more basic question: what is an earthquake? Perhaps surprisingly, even today different seismologists will give you different answers. Reading the word as literally descriptive—"earth" plus "quake"—one would define the compound word in terms of shaking of the earth. The earth quakes. Alternatively, some seismologists, including the author, prefer to use "earthquake" to mean the physical process that happens in the earth that causes the ground to quake. An etymologist would point to the first definition as correct: people have been familiar with the quaking of the earth since their earliest days as a sentient species, yet only in the last century have scientists understood the processes that generate shaking. The word "earthquake" thus existed before anyone understood what a fault *is*, let alone the nature of the abrupt paroxysms that sometimes occur along them.

Continuing in an etymological vein, the first quantitative earthquake measurements were almost inevitably based on the severity of shaking at any given location following an earthquake. (The reader may note the casual reversion to the author's preferred definition of the word.) Although such *intensity scales* are somewhat tangential to the story of the Richter scale, they are an interesting story in their own right, and one that helps set the stage for the later concept of magnitude.

Intensity scales emerged on the scientific scene towards the end of the nineteenth century, when an Italian seismologist, Michele de Rossi, and a Swiss scientist, Francois Forel, independently published similar scales in 1874 and 1881, respectively. The so-called Rossi-Forel scale includes ten intensity levels ranging from barely perceptible shaking to shaking that causes catastrophic damage. By convention, intensity values are indicated with Roman numerals.

Intensity values are not determined based on readings from scientific instruments, but rather the effects of shaking on people and structures. Intensity measurements thus impress many modern scientists as "unscientific," yet sober reflection reveals this to be an overly pejorative assessment. At the lowest levels of intensity, barely perceptible shaking can be quantified by a consideration of human sensitivities. At moderate levels, the degree of shaking that will, for example, knock small objects off of shelves but cause no structural damage can be quantified by a consideration of the stability of small objects. At the highest levels one can similarly quantify—in terms of velocity or acceleration measurements—the level of shaking that will damage structures that fall into various building types. In effect, intensity measurements are based on data from instruments: the instruments are just people, china, chimneys, homes, and so forth. One might argue (fairly) that these are crude instruments: not every chimney will respond precisely the same way to a certain severity of shaking. On the other hand, every person and structure represents a potential intensity measurement following an earthquake, whereas data from seismometers are invariably far more sparse. Using Web-based tools, seismologists can now collect data from tens of thousands of locations for large and even moderate earthquakes.

The idea of characterizing the severity of earthquake shaking is such a natural concept that it arose over and over for at least a century before the Rossi-Forel scale gained widespread acceptance in the scientific community. Italian scientist Schiantarelli used a simple scheme in 1783 to characterize shaking from a Calabrian temblor that same year. A scientific publication in the Netherlands, by P.N.C. Egen, described intensity measurements in 1828. In fact, so natural is the concept of intensity that it had been invented even earlier by individuals outside of the scientific community. When the New Madrid earthquake sequence struck the midcontinent of North America between December 1811 and February 1812, two individuals independently sprang into action as armchair seismologists. In Louisville, Kentucky, engineer Jared Brooks kept a record of every earthquake, devising a six-level scale to rank them by severity. After a while he even devised systems of pendulums that swung in response to vibrations too gentle to be felt by humans, and he kept track of how often they were in motion. In Cincinnati, medical doctor Daniel Drake kept similar records and developed a similar ranking scheme. The similarity between Drake's and Brooks's scales with later scientific scales is even more remarkable when one considers the number of levels. Although both men experienced what they considered to be very strong shaking, they were several hundred miles away from where the earthquakes were centered, locations at which shaking severity would rank no higher than VI—VII on later scales with ten and then twelve levels. Hence, they very likely invented a scale with six levels because they only experienced six levels of shaking.

But what does one mean by "six levels of shaking"? One might reasonably think that the definition of a "level" is purely arbitrary. It would, after all, be quite nonsensical to say that someone developed a temperature scale with fifty degrees because they had only experienced fifty degrees of temperature. But perhaps "intensity" levels are less arbitrary: perhaps the step from one intensity level to another has a natural meaning. Perceptible earthquake shaking spans a well-known range, from the level that can barely be felt by humans to shaking that is about as strong as the force of gravity—that is, able to send unsecured objects airborne. The very highest shaking levels recorded during any earthquake hover right around the g-force of gravity: a small handful of recordings reveal higher values, but higher by at most by a factor of two. For reasons that scientists still debate, earthquakes seem to generate maximum shaking not much above the g-force of gravity—fortunately for us all.

Now consider earthquake shaking in terms of the physical quantity acceleration, expressed in terms of fraction of g-force. If one considers "barely perceptible" shaking in g-force terms, and uses this value to define the low end of a scale, and then defines subsequently higher levels as being twice the g-force level of previous levels, one finds that it takes about ten steps to make it from "barely perceptible" up to the strongest shaking the earth can generate. Intensity steps thus emerge naturally based on shaking levels that differ by a factor of two: humans are, it seems, naturally able to distinguish steps of this size. In Cincinnati and Louisville respectively, Daniel Drake and Jared Brooks experienced a range of shaking that spanned six factors of two—not the full ten factors the earth can dish up—and created six-level scales.

Within the scientific community intensity scales were developed further by Italian seismologist Mercalli, who expanded the Rossi-Forel scale to twelve steps. The scale was further modified to account specifically for different building types by a team of scientists, Medvedev, Sponheuer, and Karnick: their MSK scale remains widely used throughout Europe today, while the Modified Mercalli Intensity (MMI) scale remains widely used in the United States.

An earthquake's *magnitude* is a measure of its inherent size, an estimate of more fundamental importance in earthquake science than intensity. In modern times it is a rare earthquake study that does not employ the concept of magnitude; studies that analyze or employ the concept of intensity are relatively few and far between. However, even the advent of modern seismometry and seismological methodologies has not, and probably will not ever, rendered intensity measurements obsolete, largely for the aforementioned reason that they represent such a rich potential source of data. Intensity values are also virtually the only data that seismologists have to investigate important earthquakes that struck prior to the invention of the seismometer in the late 1800s. The coexistence of two types of earthquake measurements leads to inevitable confusions when people fail to distinguish between shaking severity and earthquake size. Many myths arose in the wake of the 1994 Northridge, California, earthquake, including one that "the earthquake was really magnitude 10 in Northridge, not 6.7, but the insurance companies suppressed this information." One can rarely trace the origins of such myths with absolute certainty, but one often finds a germ of truth at the roots of such urban legends. In this case the seismologist has strong suspicions about the nature of the germ: intensity maps depicting values of X in and near the city of Northridge. (The statement "magnitude 10 *in* Northridge" all but screams confusion between intensity and magnitude: the magnitude of any earthquake is what it is, while its intensities will span a distribution of values.)

Returning to Charles Richter, his initial responsibilities at the Seismo Lab included measuring data from seismograms so that earthquakes could be located. To do this, seismologists rely on precise measures of when P and S waves arrive at different stations and then essentially triangulate to figure out where the earthquake occurred—much the same way that one could measure the time lag between thunder and lightening in different locations and pinpoint the location of the lightening bolt. Richter's measurements were made from seismograms recorded on the Wood-Anderson seismometers that the head of the lab, Harry Wood, had codesigned.

Richter's work on locations led him to suggest that earthquakes might be compared using the levels, or amplitudes, of shaking recorded by these seismometers. And herein lies an important point: as originally defined, the Richter scale is based on the peak amplitude, or wiggle, on the seismogram from a given earthquake *recorded on a certain type of seismometer.* The last point deserves emphasis. The measurements used for magnitude calculations do not represent the actual motion or velocity of the ground: a seismometer greatly amplifies the actual motion in a rather complex, although understandable, way. The seismogram amplitudes by themselves are thus rather arbitrary. The point of the Richter scale was not to size up earthquakes in terms of their physical energy release, for example; such developments were still decades away in the 1930s. The point was to record all earthquakes on a standard type of seismometer and measure them in a way that reflected their relative sizes in a sensible way. Another point deserves emphasis: magnitudes are not assigned on a 1-to-10 scale, the way that intensities are assigned. Rather they are calculated values based on the largest amplitude of wiggles recorded on the Wood-Anderson seismometer.

The size of the wiggles on any seismogram will of course also depend on how far away an earthquake is from the seismometer that recorded it. Thus any classification scheme must include a correction for distance. For this Richter found guidance from the Japanese literature: a 1931 paper by Professor K. Wadati that suggested how distance corrections might be determined. In an apparent effort towards political correctness, some modern researchers go so far as to refer to the "Wadati-Richter magnitude scale." Wadati did not, however, take the final step of using the distance corrections to develop an actual magnitude scale.

The next stumbling block involved the very large range of earthquake sizes. A useful classification scheme for earthquake sizes, Richter realized, must use a manageable range of values, yet "the range between the largest and the smallest [sizes] seemed unmanageably large." Richter credited his colleague Gutenberg for the "natural suggestion to plot the amplitudes logarithmically." Although maligned by generations of high school students for whom mathematics is not a favorite subject, logarithms are simply a mathematical contrivance to collapse a large range of numbers down to a smaller range. In their simplest incarnation, logarithms are simply the number of factors of 10 in any given number: the "log" of 10 is 1, the log of 100 is 2, the log of 1000 is 3. The log of 500 is a bit of a nuisance to determine precisely (without a modern ten-dollar calculator), but is between 2 and 3.

Throughout his long career Richter railed against the popular impression of logarithms as beyond comprehension. In a 1970 letter to a reporter from the *Milwaukee Journal*, he wrote, "At the risk of nit-picking, I should mention that the fundamental formulation of the magnitude scale is hardly 'complex' mathematically. It is a simple use of logarithms. As you are aware, the American daily press takes fright at the word 'logarithm,' though it freely prints more technically complex terms referring to astronautics, atomic physics, and other publicized fields." Never one to mince words, or to understand that not everyone understood mathematics as easily as he did, Richter further observed that "the blame, I fear, does not really rest on the journalistic profession—but on the outrageously inferior level of instruction in elementary mathematics in most of our high schools."

The maligned logarithm collapses a big range of numbers into a much smaller range. As many readers are aware, a magnitude 4 earthquake is thirty times more energetic than a magnitude 3; a magnitude 5 is one thousand times more energetic than a magnitude 3. Energy increases by a factor of thirty rather than ten for arcane reasons related to the nature of the earthquake process. In short, the peak shaking *amplitude* increases by factors of ten from one magnitude unit to the next, but the duration and other aspects of the shaking also change as temblors get bigger, and so the energy increases by more than a factor of ten with each step. In Richter's inimitable words, "Magnitude is important because one earthquake is a hell of a sight bigger than another one." Even having recognized this from the beginning, Richter later wrote, "When the Eureka earthquake in 1932 blasted in with magnitude 6 I was flabbergasted." This relatively obscure temblor in northern California first hammered home the point to Richter himself, just how big "a hell of a sight" could be.

A high school student today can push a button and determine a logarithm of any number, whether or not he or she has any real understanding of what a logarithm is. In the 1930s the difficult numerical problem of calculating logarithms was solved with so-called log tables. Not to be confused with log cabins and such, a log table consists of rows and rows of numbers—precalculated logarithm values that scientists and engineers relied on before the advent of modern computers. Gutenberg's innocent sounding suggestion thus represented one of the tasks that would have kept Richter's attentions occupied, and energies diverted, for many long hours. Among Richter's papers in the Caltech archives are some that testify to the arduous nature of computational science in the era before modern computers were invented: hand-crafted graphs, page upon page of calculations and hand-tabulated results. Also among his "papers" one finds a "Sun Hemmi" slide rule in a sturdy leather case, the one computational instrument that Richter could turn to for assistance. Its limitations notwithstanding, the slide rule has a weight to it, a simple yet magnificent elegance of design. Slide rules such as Richter's were treasured by their owners and well cared for.

And so Richter persevered, focusing his relentless tenacity on the daunting problem at hand—and in so doing again managed to focus his energies into gainful endeavor in physical sciences rather than allowing them to fuel his neuroses. The initial passion of his early newlywed days had waned (if not cooled completely); later corporal passions, as well as significant outside interests of his later life, still lay years ahead. Richter's later colleagues believed that seismology was the all-consuming interest of his life, throughout his life: if such was ever the case, it was during the years when the development of the magnitude scale demanded his full attention.

Putting Gutenberg's suggestion to good use, Richter continued his work. The logarithm indeed collapsed his unmanageable range of numbers to a manageable one and allowed Richter to stage further attack on the problem. Reducing the range of numbers was a key step in the development of a scale, but other steps were required. In particular, one needed to look at individual earthquakes recorded at many distances to figure out the critical distance correction. Today one can turn to the seismological literature and find stacks of literature, including the author's Ph.D. thesis, addressing the issue of so-called attenuation corrections that describe how shaking diminishes, or attenuates, with distance. Richter's efforts in the 1930s were very nearly without precedent, aside from the single prior publication by Wadati.

Looking at suites of seismograms from earthquakes of different sizes, Richter made graphs of their amplitudes as a function of distance and realized that, while curves from individual quakes were higher or lower depending on earthquake size, their shapes were the same. By shifting the curves to lie on top of one another, his measurements from all of the earthquakes defined a single smooth curve that described how shaking diminishes as one moves away from an earthquake. In the obtuse parlance of physics, this says that the earth is a *linear* system: the rate at which shaking diminishes does not depend on the initial shaking level. Appealing to automotive analogy, we say that this is like a car whose rate of braking (its deceleration) does not depend on the initial speed. Thus, if at close distances one earthquake is twice as severe as another, it will remain twice as severe at all distances. (Only in recent years have seismologists learned that this linearity does not hold for the very strongest shaking levels, but such is the stuff of which opaque seismological research papers are made.)

Richter's superposition of shifted curves thus provided what he needed: a simple way to correct amplitude measurements for the distance from an earthquake to the seismometer. (Richter defined the scale so that all amplitudes were corrected to a reference distance of one hundred kilometers.) The distance measurement, meanwhile, was safely in hand thanks to the kinds of routine, and, again, tedious, analyses to locate earthquakes that Richter had been doing already.

With a manageable range of numbers and a distance correction, all that remained was construction of a sensible scale. Richter observed that many people mistakenly assumed that the scale was "based on a scale of 10"—that is, a range of values with a maximum of 10. In fact the scale has no upper limit. The fact that 10 appears to represent an upper limit on the size of earthquakes is mostly serendipitous. It was, however, necessary to calibrate the scale somehow. After all, who was to say what kind of earthquake corresponded to what size magnitude? A key consideration in this calibration came not from the high end but the low end: Richter fine-tuned the definition of the scale so that magnitude 0 would be the smallest earthquake he considered to be detectable by a seismometer. Given this definition, negative values are possible (the logarithm of a number less than 1 is negative), but Richter wisely recognized the undesirability of having to talk about or work with negative values. It stands as a testimony to Richter's acumen as an observational scientist that, notwithstanding the considerable advances made since the 1930s in seismometer design, earthquakes with negative magnitudes remain rare. Such "micro-earthquakes" are typically recorded only in very specialized environments, for example with extremely sensitive seismometers that are installed in boreholes below the earth's noisy surface.

The values that Richter's scale yields were further intended to be generally managable: 2.5–3 for an earthquake that will be lightly felt, 6 for one that can cause light to moderate damage, depending on the vulnerability of structures in the region. The largest earthquake that Richter analyzed in his 1935 paper was large shock that had struck Nevada in 1932; for this event he estimated a magnitude of 7.5. He speculated that, based on various lines of evidence, the great 1906 San Francisco had a magnitude of at least 7, and "may have been of magnitude 8.0 or perhaps larger." Richter would have understood that earthquakes such as 1906 are especially large and relatively rare events: he did not define explicitly an upper end of the scale but rather tuned the scale to yield convenient numbers. It is thus probably only coincidental that the value of 10 emerges as a reasonable estimate of the largest earthquake that can strike on Earth, but Richter likely had some sense of how big the very biggest earthquakes would be according to his scale. The largest instrumentally recorded earthquake to date was the great Chilean earthquake of 1960, with a magnitude of 9.5. Bigger events have likely happened on the planet during human history—a huge earthquake in India in 1505 being one possible candidate—but we have no seismograms that allow us to measure temblors prior to the late nineteenth century.

The magnitude values discussed above have familiar meanings today. We "know" a magnitude 6 earthquake to be a moderate but potentially damaging shock; we recognize magnitude 8 as the iconic and much feared Big One in California; we fear magnitude 10 as the ultimate doomsday event. A final point that deserves emphasis: *The meanings that these numbers possess are the meanings that Charles Richter endowed them with.* He introduced the very word *magnitude* into seismology as well. It has been said that the ultimate contribution in science is one that everyone uses in their subsequent research and nobody references. Such contributions have become such an integral part of the fabric of a scientific field that future generations of scientists forget that somebody introduced them in the first place. Richter's 1935 publication describing the magnitude scale is well referenced as scientific publications go, but if it were, as it arguably should be, referenced by every seismology paper that uses the word *magnitude*, the number of citations would be off the charts.

A discussion of very large magnitudes inevitably segues to the limitations of the Richter scale. However, before leaving this part of the story behind, one key issue remains: having succeeded in developing a scale that he considered satisfactory for his purpose (classifying the sizes of earthquakes in Southern California), what to call it? Harry Wood pointed out that the scale needed a name that would distinguish it from the intensity scale. In his 1935 paper Richter credited Wood for having suggested the term *magnitude*, but the choice to adopt it was Richter's; with his longstanding amateur interest in astronomy, one imagines the decision to have been an easy one. Readers with astronomical affinities will recognize magnitude as the word used to classify the brightness of stars, although Richter introduced it with a twist. Whereas the brightest stars are assigned the smallest magnitude values, Richter could not abide the thought that earthquakes should be similarly ranked. Fittingly enough, in making this change he essentially turned astronomy upside down. In 1935 he wrote that he had been "apologizing to astronomers for using the term magnitude, then running the scale in the opposite sense to that used in astronomy, but no better terminology or notation suggested itself." Astronomers of the early 1930s could not be reached for comment.

For most people even today, the generic "magnitude scale" is less widely recognized than is "Richter scale." And therein lies another story. In its original formulation, Richter's scale was a way to classify the relative sizes of earthquakes in Southern California—the primary region of Richter's professional interest. In the words of Frank Press, Richter remained "California-oriented" throughout his career, while, as a European, it was perhaps natural that Gutenberg "had a world view." Thus did Gutenberg naturally contribute more to the next part of the project, extending the original definition so that it could be used to classify earthquakes worldwide. At this point the scale might have become the Richter-Gutenberg scale. The term *Richter scale*, however, was destined to leave the sheltered waters of scientific discourse and become part of the popular vernacular. For the latter purposes, Richter-Gutenberg scale simply wouldn't do. Even had Richter attempted to use this term in his dealings with the Southern California media, it is highly doubtful that the ungainly name would have stuck.

The next question, of course, is whether Richter even attempted to credit Gutenberg's contribution in his dealings with the media—the filter through which the scale would be disseminated to the outside world. Frank Press does not address this question directly but does hint at the answer: "It's a difficult question," he says, "of relations between people." Richter's own interview with Henry Spall suggests a cheerful willingness to credit his colleagues for their contributions, big and small. Yet as later chapters will discuss, it also appeared to be of tremendous importance to Richter to be seen by the media and public as the leading expert on Southern California earthquakes. One might thus be inclined to suspect that Richter did little to dissuade the general public from viewing it as the Richter scale. Richter's obituary in the *Los Angeles Times* very much reinforces this suspicion, quoting a "long-time Caltech colleague" of both Richter and Gutenberg: "There is simply no question that it should be known as the Gutenberg-Richter scale," he said, "but for many, many years, Charlie did very little to emphasize Beno's role. If you wanted to think it had all been Richter's doing, that was OK with Charlie." The article goes on to observe that members of Gutenberg's family were "openly displeased by what they saw as Richter's acceptance of the public impression that the scale was his alone."

This perception of events was reinforced in Peter Hernon's 1999 novel, *8.4.* The novel—historically accurate enough for Richter to appear in a cameo appearance—describes the Kresge Lab and the Donnelly mansion, as well as Richter's penchant for nudism. It also portrays Charlie as "a real SOB" who "screwed Beno Gutenberg." A character in the novel goes so far as to parrot words from the *Times* obituary (leaving the reader with a sense of the extent of the author's research): "If you wanted to call it the Richter scale, well, that was fine with him." Considering Richter's life at more length than the snapshot provided by a newspaper obituary, one fact becomes clear: Charles Richter was a lot of things, but "SOB" was not one of them.

Where did the term *Richter scale* come from in the first place? Richter addresses this question in a 1979 interview with Ann Scheid, telling her that that he initially used "magnitude scale," and "refrained from attaching [his] personal name to it for a number of years." Richter points to Berkeley seismologist Byerly for first referring in public to the "Richter scale." A 1966 letter from Byerly provides corroboration: "It brought back memories of how, in the early 1930's, I told the Press Associations that the magnitude scale was yours and should be referred to as such. It worked. It became the 'Richter magnitude.' " Byerly's letter continues, "Now we have a later development. At the recent meeting of the Geological Society of America it was referred to as the 'Richter' (the word magnitude omitted). The statement was like this: 'The earthquake had a Richter of five.' I liked this." The later development did not, as it turn out, stick.

Richter went on in his 1979 interview to acknowledge that the term *Richter scale* "somewhat underrates Gutenberg's part in developing it for further use." The discerning ear does note the qualifier "somewhat," as well as the fact that Richter pointed to Gutenberg's extension of the original formulation but not Gutenberg's role in development of the initial scale. Still, virtually all scientists benefit from bits of advice from their colleagues: the suggestion to use logarithms—like Wood's suggestion to use the term *magnitude*—would have been a casual bit of advice in the midst of Richter's monumental—and by all accounts, essentially solo—undertaking.

One is inclined to believe that the unnamed longtime colleague knew from whence he spoke about two men he had observed firsthand for decades: that, by outward appearances, Richter did convey the impression that the original contribution had been his alone.

Then again, it *was* his contribution. Virtually everyone at the Seismo Lab (and elsewhere) regarded Gutenberg as one of the truly great seismologists of his day. In the words of colleague Clarence Allen, who worked

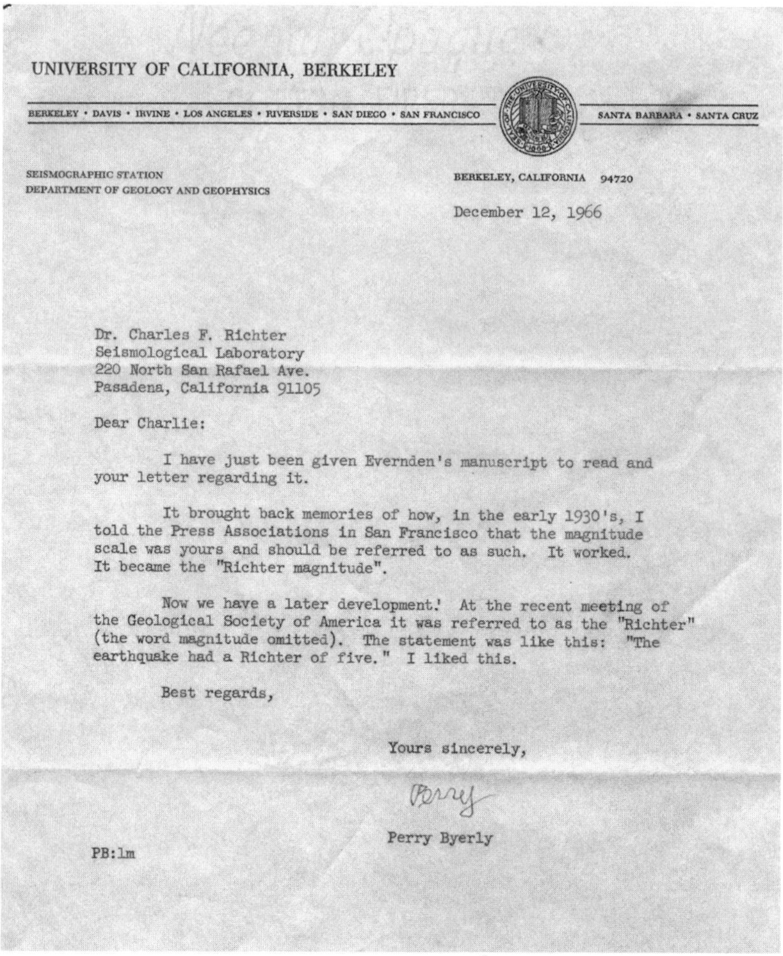

UNIVERSITY OF CALIFORNIA, BERKELEY

BERKELEY • DAVIS • IRVINE • LOS ANGELES • RIVERSIDE • SAN DIEGO • SAN FRANCISCO — SANTA BARBARA • SANTA CRUZ

SEISMOGRAPHIC STATION
DEPARTMENT OF GEOLOGY AND GEOPHYSICS

BERKELEY, CALIFORNIA 94720

December 12, 1966

Dr. Charles F. Richter
Seismological Laboratory
220 North San Rafael Ave.
Pasadena, California 91105

Dear Charlie:

I have just been given Evernden's manuscript to read and your letter regarding it.

It brought back memories of how, in the early 1930's, I told the Press Associations in San Francisco that the magnitude scale was yours and should be referred to as such. It worked. It became the "Richter magnitude".

Now we have a later development! At the recent meeting of the Geological Society of America it was referred to as the "Richter" (the word magnitude omitted). The statement was like this: "The earthquake had a Richter of five." I liked this.

Best regards,

Yours sincerely,

Perry

Perry Byerly

PB:lm

Fig. 9.1. Letter from Perry Byerly of UC Berkeley, 1969. (Caltech Institute of Technology Archives, reproduced with permission.)

closely with Richter during his final years at Caltech, "Certainly, both Benioff and Gutenberg had far more profound, fundamental contributions to seismology and plate tectonics than did Richter." A gifted observationalist to the core, nobody would have been more aware of this perception than Richter himself. And yet he was not a SOB; by all accounts he gave credit where credit was due. If he felt a unique sense of ownership about his scale, perhaps we can return the favor and give him his due. Notwithstanding the ease with which darker motivations can be found, we have no evidence that the sense of ownership was anything but rightful.

The moniker *Richter scale* earned Charles Richter an enduring measure of immortality. Few people today do not recognize his name—and perhaps fewer would recognize the name of any other seismologist, living or dead. This provides a certain measure of vexation to some scientists. "I don't mean to belittle Richter," Allen later said, "but there is no question that because of his name being associated with the magnitude scale, the public thinks he's the greatest seismologist that ever lived. Well, that's simply not true, or in keeping with the facts."

But what are the facts? Here one is led to explore a paradox that permeates research science: the individuals whom colleagues would point to as the most brilliant scientists—the greatest scientists of their day in their fields—are not necessarily the ones who make the most profound, or the most enduring, contributions. Nor are they necessarily the ones whose contributions capture the public imagination, or the ones whose contributions have the largest societal impact. Geneticist Barbara McClintock faced an uphill battle for acceptance through much of a career that focused on old-fashioned genetic experiments in the tradition of Gregor Mendel. Long after the field of genetics had moved towards sophisticated molecular biology studies, McClintock conducted classic experiments breeding corn plants to track their RNA. Recognition came only slowly; her "jumping genes" theory earned her the 1983 Nobel Prize in Physiology or Medicine at age eighty-one. In the earth sciences, Alfred Wegener faced a similar battle for respect for his early theories of continental drift; sadly he died too young to see the late-in-life vindication that might have come in the earliest days of the plate tectonics revolution.

It is perhaps not unfair to say that there are two types of brilliant scientists: brilliant inside-the-box thinkers, and brilliant outside-the-box thinkers. A scientist like Beno Gutenberg can drive a field forward full throttle by sheer force of intellect, tackling the next obvious problem, and the next after that. He had gone so far as to have important Ph.D. thesis topics prioritized, to be parceled out in turn as (sufficiently dutiful) graduate students materialized. Scientists such as Wegener, McClintock, and Richter, meanwhile, do not drive a field as fast or as relentlessly as their more organized, more focused colleagues. Yet sometimes their contributions endure.

Less famous examples abound in any field, in part because not only creativity but also serendipity play such enormous roles in scientific discovery, and serendipity strikes where it strikes. Certainly a scientist has to have the wherewithal to capitalize on the serendipitous moment. In any field of science there will inevitably be scientists who fail to appreciate the real significance of an observation or result that is staring them in the face. But by the nature of the phenomenon, the "greatest scientists of their day" do not have a monopoly on serendipitous discovery; thus do they also not have a monopoly on the greatest science of their day. Nor do they have a monopoly on contributions of enduring importance. The Richter scale itself was not considered the greatest science of its day, yet it remains one of the most fundamental contributions from the early observational era.

Moreover, if the 1935 paper can be considered a yeoman's contribution, this was no ordinary yeoman. For starters, no ordinary yeoman could have managed the monumental computational effort required to develop a magnitude scale with the rudimentary tools at hand; no ordinary yeoman would have brought such remarkable acumen to the task. The contributions in the 1935 paper go beyond the presentation of the scale itself. The paper includes a tabulation of the frequency of shocks of different magnitudes, the foundations of what would later be known as the Gutenberg-Richter relationship. Noting large earthquakes to be very much less frequent than smaller shocks but also very much larger, Richter's paper concludes, "The smaller shocks are not sufficiently frequent to contribute more than a small fraction of . . . energy. It follows that the smaller shocks do not appreciably mitigate the strains which are released in the large earthquakes, but must be regarded as minor incidents in and symptoms of the accumulation of strain." In this concluding paragraph Richter reveals prescience bordering on wizardry: the field of seismology as a whole was decades away from a rigorous proof of this conclusion, yet Richter knew the answer before he had a right to know it.

Ironically, Richter's name remains a household word even though the original formulation of the Richter's scale was all but obsolete by the end of the twentieth century. This, then, brings us to the aforementioned limitations of the scale as originally defined—a discussion that must venture into a realm of seismology that remains dark and murky for the overwhelming majority of human beings who are not seismologists.

Recall that the first step in magnitude determination is to measure the maximum shaking from a given earthquake recording at a given site, using one particular type of seismometer. Although the Wood-Anderson seismometer was a marvel for its time, the reader will not be surprised to learn that the instrument design is now recognized to have limitations, and has been supplanted by several generations of more sophisticated instruments designed to more faithfully record the full range of ground motions. To discuss this issue one must again retreat to somewhat arcane concepts, but these concepts are far more accessible by way of analogy—once again, to music. Like any piece of music, earthquake waves include a wide range of tones: high tones correspond to shaking at high frequencies (the likes of which will rattle your dishes and your nerves), low tones correspond to shaking at low frequencies (the likes of which can topple a ten-story apartment building). The Wood-Anderson seismometer is like a human ear in that it can only perceive a certain range of tones, and a fairly narrow range at that. Compared with modern seismometers, it would be fair to say that the Wood-Anderson seismometer was really rather tone deaf.

In fact, few Wood-Anderson seismometers remain in existence and even fewer remain in use. This in itself is not a critical limitation for calculations of Richter magnitudes if one has access to modern computers: nowadays seismologists can take a seismogram recorded on another seismometer and calculate what it would have been if recorded on a Wood-Anderson seismometer. One can turn once again to the world of music for analogy: the correction is like recording a piece of music played by an alto saxophone and using a computer program and speakers to replay the recording on a synthetic tenor saxophone.

The bigger problem is that big earthquakes, just like big musical instruments, produce predominantly lower tones. Essentially, then, as earthquakes get too big, Wood-Anderson seismometers no longer hear them. (This metaphor involves less artistic license than one might think: the P wave generated by an earthquake essentially *is* a sound wave traveling in the earth.) Seismologists recognized Wood-Anderson seismometers to be hard-of-hearing very soon after magnitudes began to be calculated routinely, and began to develop new generations of instruments.

It turns out to be tricky to build a small seismometer that can hear low tones—just as it is tricky to build a small speaker with good bass sound. Successive generations of seismometers had better and better hearing, leading to the development of subsequent magnitude scales based on the shaking severity of lower tones. All such scales followed Richter's lead, defined to dovetail smoothly into Richter magnitudes for small to moderate earthquakes. Thus, to a good approximation, a magnitude 3 is a 3 is a 3, no matter which of many scales one uses to determine the value. The scales may change, but the meaning of the numbers continues to be based on Richter's original definitions.

Eventually seismologists managed to build a seismometer that could record the lowest tones that earthquakes can generate. To return to the musical analogy once again one sees why this is needed: a musical recording device will differentiate between a soprano, alto, and bass saxophone only if it can capture the full range of tones that these instruments produce.

The best magnitude scale turns out to be one based on very low tones. Although obvious in retrospect, one must keep in mind that not until the second half of the twentieth century did seismologists begin to understand fully the nature of the tones produced by earthquakes.

Some of these tones turn out to be low indeed: based on theoretical considerations, by the 1950s seismologists realized that a very large earthquake could literally ring the planet like a bell, causing long-lasting vibrations with very low tones. Typical earthquake waves that people feel have periods—the length of time from trough to peak to trough—of, typically, one-tenth of a second to ten seconds. When the planet vibrates as a whole in what seismologists call "normal modes," the period can be nearly a full hour long. These odd waves nearly defy human imagination: following very large earthquakes, they can be of quite large amplitude yet still imperceptible to humans because the wave passes by so slowly. Following the devastating magnitude 9 Sumatra earthquake of December 26, 2005, seismologist Thorne Lay calculated that Sri Lanka moved up and down by a staggering nine centimeters—three and a half inches—as a consequence of these longest of earthquake waves. The languid pace of these waves, moving from peak to trough and back to peak in the span of an hour, rendered them too gentle to be felt. Still, nine centimeters is an impressive distance for an island nation to move up and down.

In 1979, Thomas Hanks and Hiroo Kanamori proposed the magnitude scale that seismologists use today. It has a name that only a scientist could love—we know it as the moment-magnitude scale—and understandably has failed to catch on among the general public. The term *moment* itself refers to the parameter that modern seismologists really use to size up earthquakes. Precisely speaking, moment measures the angular leverage on the faults that produce movement on a fault during an earthquake; imprecisely speaking, moment reflects overall energy release. Kei Aki was the first seismologist to measure seismic moment; Hanks and Kanamori later developed the magnitude scale based on this parameter.

The overall issue of magnitude has been thoroughly mangled and muddled over the years, not so much in scientists' conversations with each other but in scientists' conversations with the public. One point of lingering confusion stems from the continued use of different magnitude scales by different organizations. For many years, the National Earthquake Information Center (NEIC, N-E-I-C, not "knee-ick"), which has lead responsibility within the United States of reporting on large earthquakes worldwide, released something known as the "surface-wave magnitude."

Surface-wave magnitude is one of the scales that scientists—in particular, Beno Gutenberg—developed on route from the Richter scale to the moment-magnitude scale. Typically, surface-wave magnitudes are biased on the high side for earthquakes that are not especially deep—earthquakes such as big temblors on California's San Andreas Fault, or the devastating Izmit, Turkey, earthquake of 1999. Following the Izmit earthquake, NEIC quickly released a surface-wave magnitude of 7.9, a value that impressed the public as commensurate with the staggering images of damage beamed around the world.

The problem was, it wasn't a magnitude 7.9 earthquake. Seismologists' preferred moment-magnitude estimate soon settled to 7.4–still a very large earthquake but not on the doorstep of that creature of legend, the "great" magnitude 8 earthquake. Because of the logarithmic nature of all magnitude scales, the difference between 7.4 and 7.9 is not insubstantial. A jump of one magnitude unit, as mentioned previously, corresponds to a factor of ten in shaking levels, but a factor of about thirty in overall energy release. The media and public reacted with understandable confusion to such a substantial "down-grading" of the earthquake. And if the last few pages have veered into somewhat arcane mathematical concepts, the reader can imagine how well the explanation condenses down to the confines of a brief and clear interview with a journalist who needs a succinct answer, not a book chapter.

Journalists frustrated by talk of factors of ten and definitions of magnitude scales sometimes press for the Richter magnitude as *the* answer. And yet *the* answer would sometimes be nonsensical: the Richter magnitude of the 1994 Northridge earthquake, for example, was in the neighborhood of 5.7, a very much worse estimate than the true (moment-magnitude) value of 6.7. At an impasse between competing goals of accuracy and clarity, seismologists and journalists often skip the adjectives and resort to, simply, "magnitude."

A further substantial source of confusion stems from discussions of important earthquakes of the past: for example, the San Francisco earthquake of 1906, the Kanto, Japan, earthquake of 1923, the Alaska, or "Good Friday," earthquake of 1964. Earthquakes such as these were huge media stories in their day; they also remain very much alive in collective public memory. If one were to read the full body of literature on the San Francisco earthquake, one would find that its magnitude was initially estimated to be 8¼—the ¼ reflecting Gutenberg's view that magnitudes could not be determined, and thus should not be reported, to better than ¼ magnitude unit. Perhaps inevitably, 8¼ was rounded to 8.3 by the time Richter's *Elementary Seismology* was published in 1958. Only decades later would geophysicist Wayne Thatcher and, later, seismologist David Wald turn their attentions to available seismogram records and other data and determine the magnitude value that is generally accepted today: 7.8–7.9.

If the difference between 7.8 and 8¼ doesn't seem so bad, consider the evolving fate of the 1811–12 New Madrid earthquakes. For earthquakes of this vintage, the most reliable magnitude estimates are derived from the distribution of intensities using recent earthquakes for calibration. Between of the uncertainties associated with this calibration and the overall magnitude scale issues, scientists' estimates of the sizes of the largest New Madrid quakes has bounced from 7.3 to 8¾ to 8.1 and back to 7.4—in studies published in 1973, 1978, 1996, and 2000, respectively. Some sources still claim that the largest New Madrid earthquake, on February 7, 1812, was the largest temblor to ever strike the contiguous United States. Most seismologists—including the author—take exception with this claim, but it does not tend to inspire confidence when seismologists' best magnitude estimates bounce around like rubber balls: 7.3 one day, 8.75 the next. Even if seismologists eventually reach consensus, older magnitude estimates can live on like cockroaches in nuclear winter, kept alive not by exoskeletons but by old books and, nowadays, Web sites that refuse to die.

One can scarcely blame Richter for such confusions: they are, so to speak, not his fault. If it was an act of ego to christen—or to encourage others to christen—his "invention" with his own name, it was an effective act of ego and a remarkably useful, enduring one at that. In retrospect, one wonders if decades of confusions could have been prevented had seismologists simply used "modified Richter scale" to refer to their later, better scales. The name would not be inappropriate: for all of their increasing sophistication, every magnitude scale used today can trace its lineage directly to Charles Richter's scale. Every scale used today was designed to dovetail smoothly with Richter's scale and to yield the value for large earthquakes that the original scale would have yielded had Richter not been limited by the instrumentation of the day.

Eventually Charles Richter's measure of immortality may fade, as "Richter scale" and "Richter magnitude" fall more and more out of common usage and earthquake sizes are reported simply as "magnitudes." For now, however, the legacy remains very much alive, a source of confusion to many and a source of frustration to some—but also, in the end, a fitting tribute to the man who gave the world the earthquake magnitude scale, and in so doing, at least for a while, quieted the demons that had gotten the better of him as a younger man. Richter's life had been battered by a series of internal and external storms through his first three decades: by the end of his fourth decade his life had grown tumultuous once again, in particular his relationships with women. For the span of a single decade, perhaps a bit less, he had at last arrived at a fairly satisfactory personal solution, and had indeed made a contribution of permanent value.

CHAPTER 10

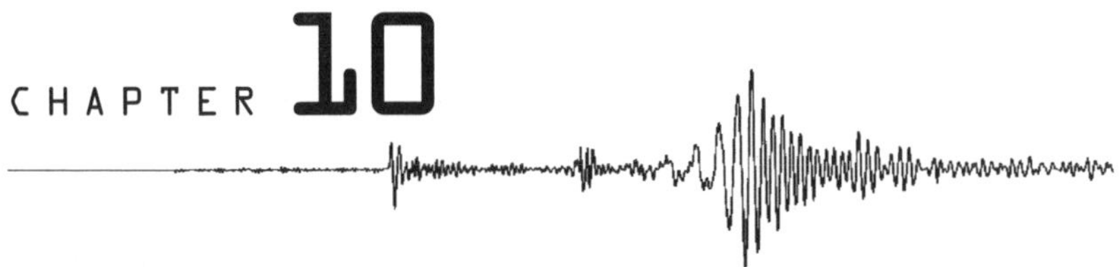

Charlie

About myself I am more sensitive; I am a rather dubious work of art.
—*Charles Richter, "Letter of Transmittal,"*
February 19, 1976

By the mid-1930s Charles Richter had settled into his career; an accidental seismologist perhaps, but a highly accomplished one nonetheless. He had established what would remain a long-term, increasingly productive collaboration with colleague Beno Gutenberg; he had developed the Richter scale; he was on his way to becoming a household world in his native Southern California and beyond. He had also, or so it seems, managed to move beyond the tribulations that had plagued his earlier years. But who was the man behind the scale? After a childhood and young adulthood of such remarkable tumult, what had he become? Richter's papers provide insights into many facets of his life; those whose paths he crossed at the Seismo Lab can also look back and paint the portrait of the man who was their colleague for decades.

Who was Charlie Richter? At a 1977 symposium held to commemorate the fiftieth anniversary of the Caltech division of which the Seismo Lab is a part, session chairman Charles Richter was introduced as a man who was "so famous, not only with us, but with the world, that he needs no introduction." Thus was the podium turned over to the man known to his Caltech colleagues as Charlie. A few weeks shy of his seventy-seventh birthday, Richter had achieved nothing less than iconic professional stature, yet were we able to view film footage of that day, many would probably be surprised by the man who took the stage.

Like many of not especially tall stature, Richter listed his height with precision: five feet, eight and one-half inches. Viewing him alongside his colleagues in Seismo Lab photographs at various times in his life, one suspects this was optimistic. Or perhaps the combined ravages of time and bad posture eroded what had at one point been a five-foot, eight-and-one-half-inch frame. At any rate, Richter was most assuredly not a physically imposing figure. He was, moreover, scarcely a dashing figure, although a single portrait of Richter as a young man managed to capture more of the intensity and promise than the eccentricities revealed by so many later photographs. Whereas colleague Beno Gutenberg had an unmistakable physical presence in spite of diminutive stature, photographs of Richter reveal a man whose appearance changed somewhat throughout his life, but never managed to lose its fundamental awkwardness.

Richter's features, including his smile, had a decidedly lopsided character. As he put on weight later in life, he developed a haphazard assortment of jowls and chins. His hair, always trimmed short, had a stubborn inclination to stick up and out towards the top on both sides, giving his appearance a mild but unfortunate clownlike quality. The ravages of time were to a large extent kind to Richter: in later photographs his hair generally looks less unruly, his disheveled grin appears less awkward and more endearing. By his sixties he looked rather like a lovable, bumbling old uncle, the likes of which are familiar on the screen in movies such as *It's a Wonderful Life*. Towards the end of his life he lost much of his padding: late photographs reveal an elderly gentleman of slight stature.

What he sounded like, therein lies another story. On tapes of the 1977 symposium Richter's voice is not what one would expect of a distinguished senior scientist of worldwide repute. His voice falters at nearly every turn, marked by a hesitant and breathy quality; high-pitched rather than deep and resonant. Stepping back through the years to another tape recording, this one from the 1950s, of a CBS news interview of both Richter and Gutenberg, we have a chance to hear Gutenberg's voice as well: accented (but not too heavily), articulate, precise, authoritative. Gutenberg does, however, reveal his lack of experience explaining science to a nonspecialist audience. At one point talking at length about such arcane matters as the temperature, rigidity, and velocity of the inner earth, the interviewer at one point interjected a clearly bemused "mm-hmm," in response.

The CBS interview highlights a general point regarding Richter's interactions with the media: some of his colleagues may have begrudged his public acclaim, but without question this "awkward teacher" was tremendously talented when it came to teaching through the popular press. His own answers to the interviewer's questions never drifted into the sea of technicalities in which Gutenberg found himself mired—along with the interviewer

Fig. 10.1. Charles Richter. (Photo courtesy of Caltech Seismological Laboratory, reprinted with permission.)

and listeners who were surely utterly lost. Gutenberg, it seems, was the type of scientist—by no means a rare breed—who either isn't interested in communicating to a broad audience, or else not able to let go of their command of a field's intricacies to explain science in understandable language.

Richter's voice as a younger man—he was in his mid-fifties at the time—was somewhat less hesitant than it would be twenty years later, especially when speaking specifically on seismology. He certainly spoke at a more appropriate level for a broad audience than did his colleague. Yet still his

Fig. 10.2. Charles Richter. (Photo courtesy of Caltech Seismological Laboratory, reprinted with permission.)

voice lacks, somewhat conspicuously, the resonance of authority one would have expected from the most famous seismologist of all time. What did Charlie Richter sound like? In the preceding paragraphs we cast our hero in the role of Uncle Billy. Now picture a television mystery in which a kindly, somewhat breathy and awkward, but very earnest and seemingly harmless librarian turns out to be the axe murderer. The voice might not be inherently sinister, but something about it hints at turmoil beneath the outwardly calm waters.

Returning to the 1977 symposium, one cannot help but note that, while the division had turned to their brightest luminary (at least in public circles) to serve as chairman of an afternoon session, they had given him a prominent but very limited role in the proceedings. This bespeaks either a lack of confidence on the organizers' part in Richter's ability to deliver a scientific lecture or an unwillingness on his part to do so. The latter seems likely: throughout his career Richter avoided scientific meetings whenever possible, not infrequently arranging to attend a meeting and present a talk but canceling at the last minute.

By age seventy-six many people are not what they once were. Still, taking the microphone Richter revealed his signature blend of wit and awkwardness. His opening remark was, "I can't refrain from referring to the date and wishing you the compliments of the season." The date was April 1. He then rambled on, quoting T. S. Eliot, "As they say, April is the cruelest month, breeding lilacs out of the dead land . . . er . . ." before introducing the first speaker, Dr. Alexander Goetz, who talked about the use of remote-sensing techniques in geologic investigations. (Richter might not have recalled the occasion some fifteen years earlier when he had mostly slept through one of the critical oral qualifying exams that Goetz had had as a Seismo Lab graduate student. For Goetz, however, it became a lasting memory. In contrast to the rest of the killers' row of Caltech luminaries on the examining committee, Richter left a lasting impression by waking up when the questioning veered to a subject that he apparently found to be of interest, asking an entirely relevant question, and, seemingly satisfied with the response, falling asleep once again.)

Goetz went on to a long career with the Jet Propulsion Laboratory and later the University of Colorado. Following his completion of his Ph.D., he was responsible for processing images collected by the *Apollo 8* and *Apollo 12* missions. He would have had a dazzling collection of images to present at the 1977 symposium. With only an audio recording and no pretty pictures, however, the biographer can only listen and fast-forward with a touch of impatience, waiting for the next chance to listen to Richter. When Richter did again take the mike, he spoke only briefly, introducing the next talk, "Hot Summers and Cold Winters on Mars." Richter did observe, in rambling fashion, that the title suggested that the talk would be a report from a manned expedition to Mars, but instead was surely a discussion of conditions that a future expedition would encounter. One suspects Richter intended this as a joke; if so, it missed the mark by virtue of hopelessly awkward delivery.

One wonders what went through Richter's mind as he sat and listened to the talk, which focused on conditions on the red planet, including results from *Viking 1* and *Viking 2*, which had landed on Mars the previous year. The stars had, after all, been a passion of Richter's since childhood. Although he never wrote of formal studies in astronomy at the undergraduate or graduate level, his childhood fascination had progressed to serious amateur star-observing during his teen years. The fascination endured and blossomed in the heady early days of space exploration. Television sets might have become a fixture in most American homes in the 1950s, but the Richter household remained without a set until 1969: if men were going to land on the moon, he remarked to relatives, he figured he should watch. When Caltech staff members Ann Freeman and Paul Roberts cleaned out his office after his death, they were surprised by the number of books and journals they found on astronomical matters.

Richter's interest in the stars extended to a long-standing passion for science fiction. In seismology circles Richter is known to have been a *Star Trek* fan. What probably few have known is the extent to which this statement was true. Here again, the well-known tidbit was the proverbial tip of the iceberg. Richter was not only a devoted fan of *Star Trek* but also, before that, of the classic science fiction magazines. Among his papers at the Caltech archives one finds long rows of archival boxes filled with his collection of magazines: *Amazing Stories*, *Other Worlds*, *Astounding Stories*, and many more, spanning nearly three decades.

The original *Star Trek* series first aired on television in fall of 1966. Originally competing against more standard family fare such as *My Three Sons*, viewers did not at first rush to boldly go where no television viewers had gone before. Within just a few months, NBC executives hinted that the series might be cancelled for low ratings. The show did, however, have its fans. NBC offices were soon flooded with thousands of letters from supporters: the few, the proud, the Trekkies. The show went on. For all of the cult status that the original series attained, it only aired for three seasons, through 1969–a grand total of seventy-nine episodes.

There is little doubt that Charlie Richter was a world-class Trekkie. In his late seventies, he kept a journal with a daily log of *Star Trek* episodes listed by title and sometimes channel. The journal ends in 1980 and includes hundreds of nearly daily entries: between his prodigious viewing habits and his remarkable memory one suspects he must have known every last line of those seventy-nine episodes by heart.

Richter's papers reveal other serious interests in addition to astronomy and science fiction. He played chess quite seriously; he listened to music, at one point compiling a list of lifelong favorites that included Brahms First symphony and the Mozart *Jupiter* finale. He and Lillian shared a passion for theater.

And then, of course, he had his writing—a lifelong passion of sufficient scope to merit a chapter all its own. Judging from the papers in the Caltech archive, he wrote two novels, one philosophy treatise, and mountains of poetry. He wrote and rewrote, finishing poems but struggling to pull together complete fiction or nonfiction manuscripts.

He did some of his writing outdoors. The mountains of California were Richter's sanctuary: a place removed from the hustle and bustle of city life but, even more importantly, from the social interactions that he found so vexing over the course of his life. He left frequently for weekend trips to the local mountains, less than an hour's drive from Pasadena: places like Chilao, Buckhorn Flats, Crystal Lake. When the magnitude 7.5 1952 Kern County earthquake struck near Tehachapi (about sixty miles from Pasadena) at 4:52 on the morning of July 21, diminished but still powerful waves rocked the Pasadena area strongly enough to rouse Richter from sleep. By his later account he then spent the next thirty seconds swearing because he had planned to leave on a hiking trip that day. As soon as he felt what he instantly recognized as a large earthquake, he knew his vacation plans had just been shot to hell.

By Betty Shor's account, the summer of 1952 was by no means an unhappy time for Richter, for whom the bounty of new earthquake data—from portable seismometers and the permanent network—made it seem like Christmas in July. Inquiries from the press and public kept lab personnel hopping as well. Shor recalled how patiently Richter answered questions from a domestic employee of an elderly woman in Pasadena. When the gentleman called for the third time to inquire of Richter, "Are you optimistic?" Richter replied with great enthusiasm, "Oh, yes, there will be many more aftershocks!" Shor adds, "That was the last phone call from that person."

Richter had no proper vacation during the summer of 1952 but did spend several weeks most other summers hiking and camping in the Sierra Nevada. Sequoia National Park remained a favorite destination for years, for reasons that the modern Southern Californian can easily understand. A modest five- to six-hour drive from Los Angeles, and less than an hour's drive east and up from the eastern edge of the Central Valley, those who journey to Sequoia, or nearby King's Canyon, find themselves in another world—a world of soaring pines and redwoods, ambling deer, clear skies, brisk waters; a world of peace and quiet broken only by the most gentle of natural sounds. He backpacked for weeks with a remarkably light load in the secondhand "Tracker" pack that he used for decades. By his account his pack weighed thirty pounds at most. Nor did he live off the land, apart from nibbling on berries he stumbled across. He did not cook at all when he backpacked alone; in fact he carried no matches. A diet of dried and canned food—combined with seven or eight miles of hiking on an average day—typically allowed him to shed eight to ten pounds of weight gained during the previous year.

Richter kept journals in which he chronicled each journey in exhaustive detail: traffic, hiking routes, food, weather conditions, people met along the way. He wrote these, along with occasional poetry, on small yellow notebook sheets, his handwriting tiny and completely inscrutable, and typed them later once he had returned home.

Richter's hobbies alone would have been enough to fill at least one lifetime, and yet from 1927 until his retirement in 1968 his responsibilities at the Seismo Lab were a full-time job and then some. Hired to perform much of the routine analysis necessary to locate earthquakes, and, later, to assess their magnitude, this sort of work remained a critical and substantial part of Richter's workday throughout his career. When asked in the CBS interview in the 1950s how he planned to spend the rest of that day, he enumerated three separate tasks: routine analysis, research projects including a revision of the magnitude scale, and preparation of his text book.

The textbook, *Elementary Seismology*, was published in 1958. Although in some respects now badly dated, the book remains on many a seismologist's bookshelf, from where it might on occasion find itself pulled out and used as a reference. The enduring value of *Elementary Seismology*, even while it predates plate tectonics theory as well as many significant late-twentieth-century seismology developments, is that it is an invaluable encyclopedia of earthquakes, one that remains useful almost half a century later. The book sold for twelve dollars in 1958; used copies of the book, whether or not they are first editions, now command a price of fifty dollars and up. The book includes basic tables and charts that are useful even today for general reference purposes, even though the information in nearly all of them has been refined by later studies. Former colleague Karen McNally marvels at the fact that while so much of Richter's early work, and that of his colleagues, has been supplanted by more recent studies with far better data and computer tools, the fundamental results have held up remarkably well. This durability stands in testimony to the extraordinary care taken in the early studies, which typically involved considerable perspiration in the form of extensive calculations done entirely by hand. Former student Shelton Alexander recalls Gutenberg's practice of never discarding any data but always basing his interpretations on those data points that he knew to be most reliable. Thus was he able to draw curves through shotgun patterns of data points—curves that, sometimes to the astonishment of later researchers, matched their own interpretations of far better data. Richter shared his colleague's deep appreciation for data.

Richter's book also includes thorough discussions of key earthquake zones around the world as well as important historic earthquakes, not only in California but also in other regions including New Zealand and India. A tremendous amount of this information is still relevant and useful. One wonders how many fifty-year-old textbooks now fetch a sum four to five times their original list price, the demand fueled not only by collectors but by new generations of seismologists eager to have their own copy for reference.

As if the tripartite demands of network analysis, research, and textbook writing weren't enough, Richter's stature as a public figure generated further demands on his time. His correspondence files include the sorts of letters one would expect a scientist of his standing to send and receive, for example collegial exchanges with Dr. Bruce Bolt, for many year's Richter's counterpart at the northern California network. One finds also other sorts of correspondence: the file labeled "Fan mail," and the one labeled, "Nut mail."

There is and perhaps will always be something about earthquakes—in particular their unpredictability—that inspires some individuals to convince themselves that they have discovered The Answer where others have failed. Some people are determined to interpret earthquakes in religious terms, others to convince themselves that they can read the tea leaves. These sorts of people wrote to Richter—quite possibly the one seismologist whose name they knew—in droves. They wrote about theories that animals can sense impending earthquakes and they wrote about much more: theories involving tides, astrology, dreams, their aches and pains, and so on. Richter's feelings on the subject of earthquake prediction were legendary, and will be reserved for a later chapter. Regarding the most off-the-wall "nut mail," Richter once observed to student John Gardner that he took special care to be kind to the writers of such letters on the grounds that "one never knew how close to the edge they might be."

Young people sent various sorts of letters to Richter, most commonly requests for information. One young writer by name of Bronwyn Fryer appears to have impressed Richter with a thoughtful and articulate request for information about faults in Southern California. Richter scribbled, "watch this boy!" at the top of the neatly typewritten letter before adding it to his files. One wonders if he ever realized that he was halfway right: *Ms.* Bronwyn Fryer went on to an illustrious career as an editor at the *Harvard Business Review.*

Yet with young people as with everything else, Richter did not mince words when someone or something stoked his ire. One young man, who shall remain nameless here, wrote to Richter in 1969 with a request for information that would have taken a great deal of time to compile. This high school freshman informed the most famous seismologist in the world, "From the data I receive I am attempting to project when the next major earthquake will occur on each of the major faults in California." He went on to add in closing, "Since I have a limited amount of time in which to complete my project, your prompt attention to this request will be appreciated." "My dear young man," Richter wrote in reply, "Your ignorance is only exceeded by your brashness and self-confidence." Richter proceeded to set the young man straight regarding the enormity of his request, and to say, "Do you not realize that the investigation of earthquakes has gone on for more than a hundred years, and if there were any regularities which could be identified, even by a young genius, 'in a limited time'(!!) they would have been worked out long ago." In closing Richter corrected the ignorance revealed by the offending writer when he requested, "a small scale model or photograph of the Richter scale."

In this modern era of Internet communications, scientists sometimes feel besieged by students and others who can track down email addresses in the blink of an eye and dash off requests almost as quickly. Now as in Richter's day, these letters run the gamut from the polite and reasonable to the impolite and obnoxious. A scientist might in his or her dreams imagine firing off a reply akin to the one that Richter wrote in 1969, but would invariably yield to discretion as the better part of valor.

People also sent Richter fan mail in droves. Some was of a professional variety: requests to be included in a compendium of biographies of famous scientists, for example. Much of it, however, was from private citizens who wrote for an autograph or to ask a question about earthquakes, or sometimes to express admiration of a gushing and feminine sort. "My Dear Dr. Richter," one such letter began in flowery and large penmanship, "How marvelous you are. You have given your life to the furtherence of science—the progress of man." The letter continued in the same vein, requesting an autograph and concluding, "With great admiration from your most ardent admirer, Sally R." One has to wonder if Richter appreciated the irony: against all odds, against all hope, in his later years he found himself on the receiving end of gushing attentions from women. His own writings, however, revealed only bemusement with the flowery, effusive expressions from women he had never met.

On a more collegial level Richter's expertise was widely recognized among seismologists not only at his home institution but also from elsewhere around the globe. Scientists wrote for his opinion on technical seismological matters as well as to ask his opinion on other scientists who were being considered for appointments or promotions. In 1965 he wrote a glowing letter of recommendation for a young seismologist named Keiiti Aki, noting in typical erudite fashion that "Kei" means "respect," or "honor," in Japanese, and "iti" means "number one." The seismological community would soon learn what Richter saw immediately: Aki went on to a brilliant career as a professor at MIT before he became the first science director of the Southern California Earthquake Center in Los Angeles, remaining at USC until his retirement in 1995.

Richter's colleagues close to home valued—and sought—his expertise as well. In one 1952 document labeled, "top secret," Richter appears to respond to a request from then—division chair Bob Sharp, who must have asked Richter for an assessment of lab activities—with an eye towards "talking points." In the single-spaced, five-page document that follows, Richter pulled no punches. He applauded Hugo Benioff's innovations with seismometer design even as he lamented Hugo's intractability on certain matters of detail. "Sometimes I still feel that Hugo has made up his mind definitely where God Almighty is still hesitating," he wrote. The letter outlines Gutenberg's rather esoteric accomplishments at length, suggesting that they might be pitched in terms such as "X-raying the earth," or "Astronomy upside down."

When he gets to the "California earthquake problem," he cautions, "Hold your hat, hobby-riding starts at this point." He then wrote of frustrations as well as accomplishments: "Almost every damn time we get an earthquake tied down instrumentally it turns out to be on a different fault than would be at first supposed." (A sentiment that rings true to an uncanny degree to modern seismological ears.) Richter offers quite a few thoughts on earthquake prediction, listing "specific results to date." First on the list: "There was original hope that little shocks would cluster along the active faults and perhaps increase in frequency as a sign of the wrath to come. Too bad, but they don't." He added, "Roughly, little shocks on little faults, all over the map, any time; big shocks on big faults." As these words make clear, Richter realized that small earthquakes occur almost everywhere in California, and neither their timing nor their locations tell us much about the big earthquakes that occur on big faults. Reading these words and many more, one understands why Sharp would have valued Richter's opinions: they were voiced with a singular mixture of acumen and candor.

Richter's files reveal that he answered a dizzying volume of letters throughout his career. He also fired off letters in response to articles or events that caught his attention or peaked his ire, for example one terse note concerning a newspaper article: "This reporter should be discharged and sentenced to hard labor for the *rest* of his life. His statements are too definite to be due to misunderstanding." Yet he responded graciously to a great many letters, some of which would have tried the patience of most research scientists.

Richter also took the time to write letters when a journalist impressed him favorably. In the 1960s he wrote a long letter to a cub reporter by the name of David Perlman, applauding Perlman's presentation at a recent Office of Emergency Preparedness meeting. Earth scientists and residents of the San Francisco Bay Area might recognize Perlman's name: in honor of his long and distinguished career, ultimately as science editor for the *San Francisco Chronicle*, in 1999 the American Geophysical Union created the David Perlman Award for excellence in science news writing. The AGU had honored Perlman previously with its Sustained Achievement Award in Science Journalism in 1997. Here again, Richter displayed his keen eye for and appreciation of talent—in journalists as well as scientists.

Betty Shor recalled another journalist who earned, and definitely received, Richter's admiration in the aftermath of the 1952 Kern County earthquake. The man had written a good story, then returned to visit the lab, asking intelligent questions and taking care to not make a nuisance of himself. One day he came in, paused for a moment, then announced, "The Indians slept outside on the ground the night before the earthquake!" Having gotten everyone's attentions with the intimation that the Indians had sensed the impending earthquake, the reporter paused to allow a crowd to gather before he continued, "The local Indians always move outside to sleep as soon as the hot summer weather begins. They having been sleeping out since mid-June." Shor adds, "Richter was delighted with [the reporter's] telling of the tale."

If Richter appreciated someone or something, he said so in no uncertain terms. If he didn't appreciate someone or something, he said so in no uncertain terms. There was method to the madness, although by outward appearances he could easily appear mercurial and unpredictable. On a personal level also, Charlie Richter could come across as a walking, breathing contradiction in terms. In the words of longtime friend and colleague Clarence Allen, who wrote a moving tribute to Richter following his death, "he could be outgoing or shy; he could be gentle and warm or abrupt and cold; and he was a man with a truly remarkable memory but, at the same time, was renownedly absentminded." Richter's absentmindedness posed a definite challenge for his right-hand woman of many years, Vi Taylor. That the two maintained such a cordial and productive working relationship is, close colleagues are inclined to believe, more of a testimony to Vi than to a supervisor who might have forgotten his head if it hadn't been attached to his shoulders.

Allen recounted two especially warm memories of his interactions with Richter. The first, reading aloud the galley proofs of *Elementary Seismology* to Richter so that he could check the errors without the excruciating exercise of reading his own manuscript yet again. (Anyone who has ever written a book is probably inclined to sympathize with Richter on this.) The exercise left Allen with the realization that, amid the immensely detailed tome Richter had buried small nuggets of humor. The second experience involved a field trip that Allen and Richter took to Baja California in early 1956, to record aftershocks in the remote region after a serious of strong earthquakes struck. "Charles may have been a genius in many ways," Allen wrote, "but he was anything but a mechanical genius, and after several bright blue flashes in his attempts to get various wires connected, I convinced him to start cooking dinner instead." How much of a genius Richter was with a saucepan, Allen does not say, but he was willing to clarify when pressed by the author: had Richter really been less dangerous with a gas camping stove than a portable generator? Allen replied diplomatically, "I should perhaps have said that I convinced him to *help* in preparing dinner." (One begins to gain a measure of understanding of why Richter ate only uncooked food during his later, solo backpacking trips.)

Fig. 10.3. Clarence Allen. (Photo courtesy of Caltech Seismological Laboratory, reprinted with permission.)

But the more memorable part of the trip came later, when a poor local farmer wandered over and ended up engaged in a long and lively conversation with Richter, in Spanish. Allen had not even been aware that Richter spoke Spanish—in fact, that he had reading and speaking knowledge of a half-dozen languages. Richter did not visit the area again, but Allen returned several times in the next few years, and Richter and his friend Manuel would ask about one another.

Allen talked about Richter's foibles, including his unpredictable temper. "It was not always easy," Allen wrote, "to judge what kinds of things would intrigue him, and what kinds of things would inflame him." Hiroo Kanamori, a Caltech postdoc in 1966–67 who was later hired as a professor in 1972, also recalled Richter's mercurial temperament. It seemed to depend on Richter's mood at the time, whether or not he would talk to any given person at any given time—but regardless of mood, one could often find him at work in the "measuring room," talking to himself. In the words of former Seismo Lab student Nafi Toksöz, Richter could "very happily carry

Fig. 10.4. Hiroo Kanamori. (Photo courtesy of Caltech Seismological Laboratory, reprinted with permission.)

on a conversation all by himself." Another student, Ta-Liang Teng, recalls having been most impressed by Richter's close mentoring relationship with student John Gardner: when Teng walked past the measurement room, Richter could often be seen talking to Gardner, who responded with polite nods and murmurs. Teng finally realized that the Gardner had a problem with his diaphragm that caused him to make involuntary sounds, and Richter's apparent repartee wasn't directed at John or anybody else. This habit of his could be unnerving: it was never quite clear when, if ever, Richter's remarks were directed to somebody other than himself. To the young daughter of one colleague Richter was more than unnerving: he was scary.

Seismo Lab staff and young children tended to tiptoe around Richter, afraid to say something that might earn his wrath. To some he came across as a cantankerous old buzzard; others saw him as simply unpredictable. Kanamori recalls a time when several seismologists at the lab discussed one of the appendices in Richter's 1958 book: almost all of the magnitude estimates seemed too high, and rather conspicuously so. Nobody dared ask Richter about this except for, by Kanamori's recollection, Robert Geller. Yet Richter answered the question both calmly and honestly: the magnitudes in the table were indeed in error. Richter had gotten a list from Gutenberg and converted one kind of magnitude ("body wave magnitudes") to another, a conversion that nearly always raises the initial estimates, not realizing that Gutenberg had already done the conversion himself. It would have been an embarrassing admission on Richter's part, to own up to such a rookie mistake, and yet he responded to the student's question calmly and with candor. Many things were difficult for Richter, but owning up to his mistakes, it seems, was not one of them.

Clarence Allen did describe tremendous personal sensitivities on Richter's part: matters pertaining to him personally, as opposed to his science, especially if he thought that people were making fun of him. Allen's comments would have, as it turns out, come as no surprise whatsoever to Richter himself. Consider the following poem, dated February 19, 1967, that appears to have been sent along with a collection of verse that he turned over to a writing teacher or friend for evaluation:

Letter of Transmittal

These hopeful sketches are now yours to see,
To cover with approval or rejection,
Or mark where they have fallen from perfection
And send them back for second draft to me.

Whether I draw an angel or a tree,
My lines are always ready for correction,
Which does not drop me into deep dejection;
I pride myself on equanimity.

About myself I am more sensitive;
I am a rather dubious work of art,
But I cannot make up my mind to live
Calmly and holding out a place apart.

Whatever gift that I consent to give
Tears out a little root inside my heart.

Ever the astute observationalist, Richter was well aware of his own sensitivities. Yet this and other mercurial aspects of Richter's personality, which we will see may have in fact been manifestations of a neurological disorder, were insubstantial details in Allen's estimation. In closing he offered a quote that a longtime associate of Richter's had offered to him: "You know, I sort of loved the man." (Although Richter never had opportunity to write a personal testimonial about Allen, without question the respect was mutual: Richter considered his younger colleague to have been the cream of the crop among the lab's early graduate students.)

Hiroo Kanamori, who arrived at Caltech in the waning years of Richter's career, further described the admiration that he had felt at times amid his overall sense of bemusement. Kanamori first met Richter not at Caltech but rather during Richter's 1959–60 visit to Japan, during which time he presented seminars on earthquakes in Japan. Richter displayed some familiarity with the Japanese language (in fact he had labored at some length to learn it) and even more familiarity with Japanese earthquakes. Kanamori, now well known for a remarkable depth and breadth of contributions to earthquake science, had begun his research career developing and deploying sophisticated instruments to measure gravity; he credits Richter with having, ironically, provided his first real introduction to earthquakes in his native Japan.

Many of Richter's closest colleagues generally felt a measure of admiration—albeit often tinged with ambivalence—towards their iconoclastic colleague. Those who knew him best felt a measure of affection as well. Nafi Toksöz recalls his former mentor with warmth and appreciation. In the intense academic environment—one that tested the mettle of any young student—Richter was a source of tremendous support, a rare and much appreciated voice of encouragement as Toksöz faced hurdles such as big exams. Toksöz went on to his own long career in seismology at that other institute of technology: MIT. Richter's last Ph.D. student, John Gardner, also recalls Richter as a supportive advisor. By the time Gardner arrived at the Seismo Lab in 1958 the computer era was just barely under way. Gardner was among the Seismo Lab's earliest programmers, back in the days when a modest computer was the size of a small house refrigerator and an IBM Model C typewriter served as the user interface. (The IBM Selectric typewriter, with its revolutionary ball element, remained a few years away at the time.) Scientific computing proved to be a career-long detour for Gardner: after receiving his degree he was quickly snapped up by Leon Knopoff, who had just purchased the first computer for the earth science department at UCLA.

Tom Heaton, who arrived at the Seismo Lab as a graduate student towards the end of Richter's career in the 1970s, described his first encounter of one of the founding fathers of seismology thus: "My first impression of Richter was that I had met a hobbit." Short in stature, "frumpy with bushy eyebrows," rather rounded in his later years, and inclined towards a mischievous sense of humor—one can understand how Tolkienesque associations would have come to mind. Heaton described an occasion when he and Richter were guests on a talk radio show. One female caller phoned with a breathless appeal: "Oh Dr. Richter, I'm so afraid of earthquakes.

What should I do?" In Heaton's words, "Without any hesitation, Richter replied, 'Why don't you get the hell out of the state?' "

The FCC would not have been amused. One suspects the unfortunate lady caller was not, either. The reply was, however, vintage Richter. So, too, were his words in a 1964 article, "Our Earthquake Risk—Facts and Non-Facts," that he published in the Caltech magazine *Engineering and Science.* Near the beginning of the mostly staid article one finds a short paragraph: "Just as some persons have a fixed neurotic fear of cats, others have an excessive and unreasonable fear of earthquakes. They should not try to live in California. More generally, people who are not prepared to act sensibly in a general calamity, but think only of their personal safety and personal interests, are not good citizens; they are not wanted in California." As Clarence Allen put it, "Charles made no pretense of being a diplomat or a politician, and in things scientific, he said what he meant bluntly and precisely—whether it was with regard to earthquake prediction, the safety of high-rise buildings, or the mental competency of selected newspaper reporters!"

Any number of things left Richter both frustrated and infuriated: incompetent reporters, incompetent citizens, and needless earthquake risk among them. On these matters he spoke directly from the heart, and sometimes with venom. Having had his home demolished to make way for the 210 freeway, Richter sometimes told people quite candidly that he was not inclined to encourage anyone to move to, or stay in, California, which clearly had far too many people already.

To a casual observer his reply to the hapless female radio caller probably appeared equally blunt—even venomous. Tom Heaton, however, who by that time had come to know Richter, the reply revealed not malice but instead candor tinged with Richter's sometimes devilish sense of humor. Longtime lab technician and later senior instrument specialist David Johnson recounted Richter's wry sense of humor as well as his many contradictions. Richter's love of hiking was renowned, but he also loved to walk; his journals reveal that even into his seventies he continued to walk, at least occasionally, from the lab to his home in Altadena, a distance of about five miles. Johnson once watched him walk along the road leading to the Seismo Lab, right foot on the curb, left foot in the gutter, and so forth as he ambled along in a decidedly lopsided gate—a bit of personal amusement after Bilbo Baggins's heart.

Further illumination of Richter's sense of humor is provided by a 1952 memo that he chose to keep among his papers, from colleague Albert Engel:

> *Memo to*: The Staff *Subject*: Effluvium
> *From*: A bleary-eyed colleague who has just dragged through a 388-page biffle euphemistically called a Ph.D. thesis. It's still not clear to me whether the author wants to be a scientist or a third-rate Thomas Wolfe.
>
> I propose that any applicant who submits a thesis longer than 200 pages automatically flunks, with no recourse; and that any staff man who accepts the pile be required to dispose of it.
>
> The revised edition of Roberts Rules states that unless more than two-thirds of the voting body are outraged by the proposal, it is automatically adopted. Owners of stock in Paper-mate pens and St. Regis Paper are barred from voting.

Richter's thoughts on this memo were not recorded. One can, however, only imagine that a memo such as this that met with his disapproval would have ended up in the wastepaper basket rather than in his files for safekeeping.

Richter also kept a long letter that he penned himself about colleague Hugo Benioff, whom Richter clearly regarded with tremendous warmth and respect. "His associates find nothing difficult about his last name," Richter wrote, "but judging from the mail, others do." The letter continues, "once a sweet voice on the telephone inquired, 'Is this Dr. Benny's office?' A less forgivable mistake is 'Huge Benioff.' (Height 5 feet 10 3/8 inches; weight, 174. Eyes, grey-green. Principal allergies: wheat, cat fur, dog hair—not the hair of the dog that bit, however.)"

David Johnson, whose interactions with Richter focused on technical matters but were congenial nonetheless, recalled that Richter "either liked you or he didn't." Some people, including many students, simply seemed to rub him the wrong way, and he couldn't be bothered to deal with them. And yet, while one gets the sense that some scientists can't be bothered to deal with anyone except other scientists, Richter's interactions with people could by no means be explained thusly. One wonders, however, if Richter really "couldn't be bothered to deal with some people," or perhaps if some people simply flummoxed him more than others.

In many ways Richter didn't quite know what to make of the world—and the world tended not to know what to make of him, either. To Seismo Lab technicians who saw him almost daily, Richter could be pleasant but inscrutable in his casual communications. To those who saw him in public appearances, either in person or over the airwaves, he could sound authoritative or he could sound irreverent, or worse. To journalists with whom he frequently interacted during his long career, he could be especially gracious or he could be scathing. To colleagues who did not know him well, and even sometimes to close colleagues, he appeared mercurial at best.

Some colleagues appreciated Richter's charms; others, not so much. One arrives at the latter conclusion not by direct evidence but rather by indirect inference, for example in the form of intimations that Richter was more famous than he "deserved to be" by virtue of underplaying Beno Gutenberg's contribution to the magnitude scale and, perhaps, grandstanding for the media. The element of resentment appeared to grow towards the end of Richter's career and life, as his fame evolved into nothing less than iconic stature—nearly universal name recognition, not only in California but also in far corners of the world. Other factors may have contributed to the backlash as well. By the end of his career, Richter's earliest colleagues—Beno Gutenberg, Harry Wood, Hugo Benioff, and others—had left the scene, replaced by scientists who knew Richter less well, and had heard only secondhand accounts of his contributions to his long and productive collaboration with Gutenberg. Hertha Gutenberg, who outlived not only her husband but also Richter, supported her husband as passionately after his death as she had during his life; she surely added her own spin to those accounts.

Some of the lore may have been rooted in kindness. At Richter's memorial service in 1985 several speakers, noting Hertha's presence, made of point of emphasizing the joint contribution made by Richter and Gutenberg during their many years of collaboration. Don Anderson described the "Gutenberg-Richter magnitude scale" that was applied to earthquakes around the world.

"There is simply no question that it should be known as the Gutenberg-Richter scale." The colleague who made this remark following Richter's death remained unnamed in the *Los Angeles Times* obituary, but one thing is clear: he could not have been at the Seismo Lab in the 1930s when Richter first (purportedly) developed the scale, as nobody else from that era was still alive in 1985. Richter, initially hired as an assistant, had been the Seismo Lab's most junior scientist when it began operations in the late 1920s. Richter would later write or relay his remembrances about his earliest colleagues: Gutenberg, Benioff, Wood. They never had the chance to do the same about him. The lore about those early years would be written later. In fact, in a 1958 letter of recommendation for Richter's Fulbright application, Gutenberg wrote, "The applicant's professional qualifications are outstanding." He went on to say, "He has introduced the magnitude scale which carries his name into seismology."

A 1989 letter to the *New York Times* succinctly summarized the other school of thought about the development of the magnitude scale. Describing himself as a member of Richter's Pasadena congregation, Robert Kaufman parroted the line that "it's really the Gutenberg-Richter scale," adding that Gutenberg did not talk to the press because he was hard of hearing, instead sending Richter, who "enjoyed the attention." In closing, the letter mentioned Gutenberg's close relationship with Albert Einstein: two truly great men of science. The fact that Richter was devout only about his atheism—nothing in his papers suggests that he ever belonged to any congregation (unless one counts nudist groups)—leads one to wonder exactly whom Kaufman had known, and exactly what motivations led him to take pen in hand.

Wherever the lore came from, and whoever had contributed to it, it had clearly become well established even before Richter's death, and perhaps nowhere more than within the corridors of the Seismo Lab itself. One Seismo Lab staff member who first met Richter in the 1970s, after his retirement, remarked tellingly, "Unlike most people, I liked him."

Richter's colleagues may have had different opinions of the man, but they tended to agree on one thing: they all saw a man for whom seismology was the one consuming passion in life. "He was absolutely dedicated to his science," Allen wrote, "almost to the exclusion of everything else." Toksoz also recalled Richter's long hours in the lab: many early mornings and late evenings, on weekends as well as during the week. It was a well-known part of the lore at Caltech and beyond that Richter had installed a seismometer in his living room. His colleagues saw all of these things, but they saw virtually nothing of his private life, and knew virtually nothing about his outside interests. Few even knew that he had a stepson; even fewer knew he had a sister.

Those colleagues who knew Richter best did begin to see and appreciate the warmth beneath the peculiar and sometimes prickly exterior, the humor beneath the sensitivities, the compassion beneath the obliviousness. Those who knew Charles Frances Richter well knew him as Charlie—not Charles and certainly not Dr. Richter—a moniker that in itself conveys an unmistakable aura of warmth. Those at the Seismo Lab who were closest to him did develop an underlying and abiding sense of affection as well as respect for their enigmatic colleague and friend. Yet even Richter's closest later colleagues may have harbored ambivalent feelings about the extent of his fame in a public area; certainly they knew him as famously mercurial, even explosive. And while it might be said that to know Charlie Richter was to love him, to love him was not necessarily to understand him.

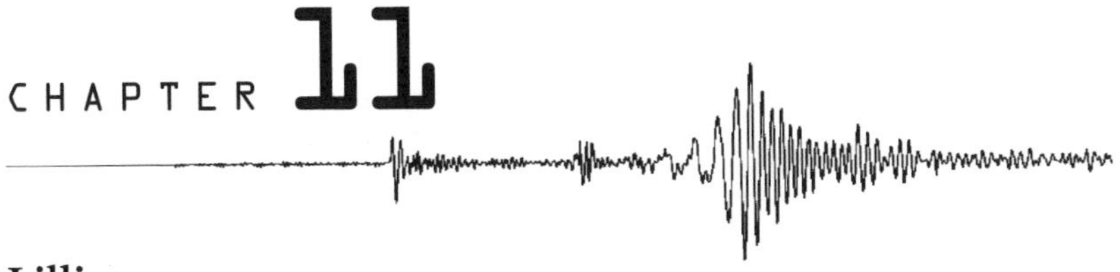

Lillian

We did not feel that we were too dependent on each other—Lillian was a very independent person.
—*Charles Richter, 1970*

RICHTER might have remained inscrutable to even those colleagues who knew him best, yet his closest professional peers knew him even less well than they imagined. In fact this "singularly focused" scientist had chapters in his life that none of his colleagues knew existed. Those chapters begin with a small handful of women around whom his personal life revolved. According to the official record of Charles Richter's life, three women played significant roles in his life: his mother, sister, and wife—respectively, Lillian, Margaret, and Lillian. The official record is, as it turns out, incomplete in this regard. These complications notwithstanding, these three women formed the fixed constellation of Richter's universe throughout his life. Two we have already met: his mother, Lillian Anna Richter, and his sister, Margaret Rose Richter. This chapter focuses on the second Lillian in Richter's life: Lillian Brand Richter.

As we have seen, Richter met the second Lillian in his life in 1927 and married her the following year. His sister had returned to Los Angeles and was living with her mother when Charles and Lillian were married. By Richter's account it had been a mistake to return home with Lillian. He describes his wife as having been "a little wild" at the time. We have only limited information about the nature of Lillian's early wild ways. Richter's papers include little material related to his wife—her writing and outside interests in particular, but general aspects of her life as well. She did have a sister, Ethel, who had been born two years later than Lillian, in October 1901, and who died in April 1976. Ethel, whose married name was Walport, had two children, a daughter named Dorothy born in 1927 (later Dorothy

Crouse) and a son Bruce born in 1932. Dorothy and Bruce grew up in Van Nuys: their mother was employed as a dietician with a military hospital. Bruce and Dorothy visited the Richters' home frequently during their childhoods: their recollections fill in many of the blanks about the woman they knew as Aunt Lil. Bruce Walport recalls a strikingly tall woman, perhaps five feet, nine inches: not a raving beauty but with high cheekbones and lovely auburn hair. She "had a certain presence," he recalled, the type of woman who would walk into a crowded room and, without saying a word, bring conversation to a stop. Crouse describes both her mother and her Aunt Lil as having been pretty. Photographs of Lillian at Richter's 1970 retirement party reveal a woman who was taller than her husband, her hair permed into salt-and-pepper curls, her figure not entirely trim. Her visage on this occasion appears more stern than striking. Richter's colleague Bob Taylor recalls Lillian as having been in good spirits that day: she had wanted him to retire for some time, but he had resisted, mightily. Taylor recalls the day that Richter received a letter from Caltech, suggesting that he retire: "He said, 'they can't do that,' and angrily tore up the letter."

Lillian and Ethel's mother died when the girls were very young. Their father remarried a woman by the name of Emma Gish, reportedly a distant relative of actress Lillian Gish. Emma brought her strict Nazarene views and lifestyle into the Brand household and her stepdaughters' lives. While Lillian would later share, and even laugh about, her "wicked stepmother" stories with younger relatives, great-niece Kathy Walport Haag observes wistfully that what passed for strict parenting in the early twentieth century would today be considered child abuse. Here we arrive at a measure of understanding of Lillian's wild ways: when she broke free from the stepmother's sphere of influence, she really broke free. In the 1920 census Lillian is listed as one of six lodgers sharing a residence at 1553 Orange Street in Los Angeles: four men who listed their occupations as "commercial travelers" (salesmen, apparently) and, in addition to Lillian, another young woman, Marguerite A. Barsot, age eighteen. Lillian and Marguerite both listed "none" under occupation; it is thus unclear how they supported themselves. By this time, however, Lillian had begun studying at the Southern Branch of the University of California, which later became the University of California at Los Angeles. She later transferred to the University of California at Berkeley, where she earned a bachelor's degree in English in 1921.

Later in her life Barsot, who never married, spent time in the Philippines, eventually returning to California and working for many years as a librarian in the desert community of Lancaster. Dorothy Crouse recalls

Fig. 11.1. Lillian (*right*) and sister Ethel (*left*), circa 1906. (Photo courtesy of Laurie Walport.)

Fig. 11.2. Lillian Brand with a group of friends, circa 1920. Lillian is standing third from left, in back. (Photo courtesy of Laurie Walport.)

Marguerite and Lillian as having been close their entire lives, "like sisters." Marguerite was Richter's friend as well: her address pops up in his date books in the 1960s, located in a tidy, elegant small apartment complex on Oakland Avenue in Pasadena, about a mile from the Caltech campus.

Some aspects of Lillian's "wild ways" start to become clear. By the time she met Richter she had already been married, had a child, and separated from her first husband, a fellow student at Berkeley. That she had a son by her first marriage emerges as one of the many surprising bits of information in Richter's papers. Only a few snippets of information about Reginald Floyer Saunders Jr. appear in his papers, but in 1949 Richter noted that he and his stepson, "Butch," were close ("much attached to each other"), and that the young man was at that time a senior at UC Berkeley. Richter's vacation journals in later years reveal that he and Butch occasionally took hiking trips together, and Bruce Walport described his cousin as a frequent visitor to the Richters' home. Charles and Butch spent many long hours playing chess. In a letter to a friend dated November 1956, Richter remarked that Butch was at UCLA studying for a masters degree in chemistry. UCLA records, however, reveal that by early 1957 the young man had withdrawn without completing a degree.

Fig. 11.3. Wedding portrait of Lillian Brand and Reginald Saunders, circa 1922. (Photo courtesy of Laurie Walport.)

The mystery of Saunders's whereabouts is solved, but not happily, with a visit to the Richters' gravesite in Altadena. Just to the left of the headstone that marks the joint gravesite of Charles and Lillian one finds a headstone that reads, "REGINALD F. SAUNDERS JR. 1925–1957. BELOVED SON." Cemetery records reveal that he was buried at the end of May of the latter year.

Fig. 11.4. Lillian (*left*) and sister Ethel (*right*) with Ethel's daughter Dorothy (*left*) and Lillian's son "Butch" (*right*), circa 1930. (Photo courtesy of Laurie Walport.)

Richter's papers, including his date books, are mute on the subject. To the casual observer, nothing about this grave marker links it to the one to its immediate right. Many of Richter's closest colleagues were unaware that the stepson even existed.

One of Richter's former colleagues, Frank Press, did recall the cause of the young man's death: suicide. Bruce Walport confirms the recollection. Although newspaper reports decorously described the cause of death as a heart attack, the young man had been despondent for months over the breakup of a romantic relationship. According to her niece, the wind never would return fully to Lillian's sails following the loss of her only child. Although she rebounded from the throes of her immediate grief, she was never again quite as vibrant as she had been in her earlier years. The death came as a tremendous blow to Charles as well. As he would do again fifteen years later when Lillian died, Richter responded to profound personal loss by withdrawing from the outside world. His papers include markedly little business correspondence during the months following both deaths.

Neither Richter nor Lillian was the other's first love, but Richter's prior relationships were apparently far more limited than Lillian's. Richter wrote very briefly that his first love affair had been at the age of twenty-one. Whether or not this was a love affair in the modern sense of the phrase, Richter was clearly, if not a virgin, almost certainly an extremely inexperienced young man of twenty-seven at the time he met his future wife.

Lillian and Charles spent their early married years in the family home on Bronson Avenue. Margaret's return from Arkansas apparently inspired them to pull up stakes. They spent about ten months in an apartment, "then borrowed money and financed the building of our present house on a shoestring." They moved into this house, on 1820 Kenneth Way in Pasadena, in August 1936. Its shoestring financing notwithstanding, the house sat on a large lot and was designed by noted architect Richard Joseph Neutra, considered one of the most influential modern architects of the twentieth century. Neutra had moved to Los Angeles in the 1920s. By the time he died there in 1970 he had left his mark on a cultural landscape that initially left him underwhelmed. His elegant modernistic designs reflected a departure from standard architectural convention. The Richters' house was showcased in a 1937 issue of the *Architectural Forum*: photographs show a structure of modest size, with two small bedrooms and a study, and an elegant, Spartan design. Much of the furniture was built in, a feature touted as "reduc[ing] housekeeping to a minimum and contribut[ing] to a restful atmosphere created by simplified construction and close harmony with the natural environment." The photographs suggest that, while the house might have been restful, it did not look particularly comfortable.

By Richter's account, Lillian had been the one with the interest in architecture. "In this matter," he said in 1979, "I was pretty well led, because it didn't matter as much to me as it did to her. After all, it was her house." Not fond of societal conventions, Lillian would have been understandably drawn to Neutra's avant-garde style. The minimalist style also suited her

Fig. 11.5. House designed by Richard Neutra for Charles and Lillian Richter. (Reproduced from *Architectural Forum*, 1937.)

aesthetic tastes. Laurie Walport, daughter of Richter's nephew Bruce Walport, recalled that the Richter's house was always elegant and tidy, inside and out. "A place for everything and everything in its place," Aunt Lil had been fond of saying. She did not abide clutter and had no use for the decorative knickknacks of which women are supposed to be fond. "There was a second-hand personal acquaintance [with Neutra]," Richter further described, "so [their arrangements were] set up on a rather cordial and effective basis." He might have described the place as "her house," but by the recollection of longtime colleague Clarence Allen, Richter loved the place as well.

It thus came as a great blow and enormous trial when the path of the 210 Foothill Freeway crossed paths with Richter's small but architecturally significant and much-loved home. In 1969 the State of California laid claim to the house by eminent domain, a trial under any circumstances and one that was exacerbated by what Richter saw as inadequate recognition of the home's architectural significance and value. A brief entry in Richter's daily diary on August 15, 1969, reads, "Bad news: Jury accepted state approval." The follow day's entry includes a similarly terse but telling entry: "L and I both upset getting home." This was, in Allen's words, "an unhappy episode

Fig. 11.6. Inside view of Richters' house. (Reproduced from *Architectural Forum*, 1937.)

in Charlie's life." On October 24, 1969, the diary noted a "last book group meeting at 1820 Kenneth Way." (Both Charles and Lillian were longtime members of a small book group that included writers, teachers, and readers from the area.)

A house can also be found today at Richter's old address on Kenneth Way in Pasadena, a tidy but architecturally insignificant one-story California ranch, now brightly painted in a mixture of brick red and earth tones. No longer a through-going street, Kenneth Way makes a 180-degree U-turn as it approaches the freeway. The house at 1820 is at the far end of the U, its lot bisected diagonally by a not entirely effective sound wall adjacent to the freeway.

In 1969 Richter and Lillian moved to a cottage on Villa Zanita in Altadena, where they remained until—and he remained after—her death in 1972. Villa Zanita was a rustic oasis: a short private road, still unpaved in 2005, lined with modest one-story houses and many trees. In 1969, part of the area around the lane was still undeveloped. The house offered several amenities that endeared it to Richter's heart: proximity to two local hiking trails, and shallow drawers in the kitchen that turned out to be just the right size to hold paper records from his seismometer. In fact, with two other offers for the property already on the table, Lillian made a full-price offer on the house while Richter was hiking in the Sierra and unreachable

Fig. 11.7. House on Villa Zanita in Altadena where Charles and Lillian Richter moved in 1969 and where Richter remained until shortly before his death in 1985. (Photo by author.)

by phone. Realtor Elaine Simmons expressed concern that he might object to the unilateral decision; Lillian replied airily, "Oh, I know what he likes." By Simmons's account, he was indeed pleased, first to be able to store his seismograms without rolling them up, and second to be able to take his regular early morning walks on hiking trails instead of city streets.

The house at 594—the Richters' former domicile—still stands, a modest wooden structure with a large wall of windows in front. In moving to Villa Zanita, Richter got about as close as he could get to the mountains he so deeply loved. Around the time of his move he also expressed an interest in getting as close as he could to the Sierra Madre Fault, which skirts the southern flanks of the San Gabriel Mountains—so that he could be close to the action in case something happened. One is not altogether sure he was joking.

In 1936, Kenneth Way would have been tucked into a fairly remote—and not terribly affluent—corner of Pasadena, just east of the Arroyo Seco, in which the Rose Bowl is situated. The house is a good five miles from the Caltech campus—a blissfully short commuting distance nowadays but presumably less trivial in 1936. However the house was less than a mile from the Kresge Lab, where Richter spent most of his time. Richter frequently walked to and from the lab when he lived on Kenneth Way; he continued to walk to the lab, at least occasionally, even after the move to Altadena.

Charles and Lillian built their first home in the midst of the Great Depression, but by Richter's account the individuals associated with the Seismo Lab and Caltech mostly soldiered on. He quoted a colleague's remark: "We took a 10% cut in salary but the cost of living went down 20%." Clearly, however, these lean years: by Richter's account he "wasn't drawing very much money from the Carnegie Institution," and his salary did not rise for many years after the lab became part of Caltech in 1936. Salaries at the lab in fact remained fixed from around 1929 until 1946. In 1943 Richter's salary was $3,800 per year, not a pittance at a time when the average American salary was less than half as much—although only about half of the salary that Gutenberg had negotiated for himself in 1930.

But if Richter's financial circumstances remained in a holding pattern during the 1930s, Lillian expanded his horizons in ways he probably never would have imagined as a younger man. In 1935 the couple became members of the Fraternity Elysia, a nudist organization headquartered at 9804 La Tuna Canyon, initially run by Hobart and Lura Glassey and their business partner, Pete McConville. Organized nudism had been imported into the United States from Germany less than a decade earlier. The original Fraternity Elysia, located in the hills near Lake Elsinore, was only the third organized nudist camp in the country. And here we arrive at one of the few personal tidbits that is generally known in seismology circles: "He was a nudist, you know," seismologists sometimes observe with a grin. The grin does not tend to diminish when one recalls photographs of Richter, who grew from awkward and thin as a young man to awkward and less thin by middle age. Nudists will tell you, however, that nudism is not about sex or exhibitionism, but rather about an acceptance of oneself and a closeness with nature. Either way, within the staid mid-twentieth-century corridors of Caltech, many regarded a nudist seismologist with unease if not distaste.

Nudism was for Richter very much about closeness with humanity, although not (for the most part) closeness of a sexual nature. He wrote that Lillian was a "more 'natural' nudist" than he was. In his words, she was, "an open-air, somatotonic type of personality, and rejoices in fresh air, sun, and healthy exercise." (A somatotonic individual—one learns with assistance from Google, is active, courageous, and competitive. In astrology circles the personality type is considered to be intermediate between Mars and Jupiter types. Enough said.) For Richter himself, however, the appeal lay elsewhere. "I might say," he wrote, "that I had never known what real friendship was; I had never had any intimate friends—until we joined the Glassey group." Richter did not say how he and Lillian came to be acquainted with the group. They may have attended one of the talks that Hobart Glassey gave at the Clifton Cafeteria, a restaurant in downtown Los Angeles. Or perhaps Lillian viewed the short film *Elysia*, produced in 1933 by Bryan Foy of Foy Productions to promote nudism. Glassey also spoke under the auspices of the Los Angeles Forum, at the time a key sponsor of cultural and educational events in the city, and at the landmark Pantages Theater in Hollywood when *Elysia* premiered. In the movie, a journalist visits the camp to interview its founder, played by Glassey himself. The film reveals Glassey to have been of short stature but lithe, whippet-thin, intense, and articulate. The conceit of the interview allows him to elaborate at length about the benefits of social nudism: the practice is described as emotionally wholesome as well as enormously beneficial physiologically, a cure for everything from neuroses to rickets. To an open-minded young woman who, as we will hear shortly, suffered with serious health problems throughout much of her life, these claims would have had obvious appeal.

Glassey chronicled the founding of the camp in several articles that he published in *The Nudist*, an early nudist magazine: originally situated on leased land, by 1934 Glassey and his partners had purchased 250 remote acres near the town of Elsinore. By January 1935 the amenities included eight cabins as well as a dining hall, kitchen, and dormitory. Recreational facilities included two volleyball courts. Glassey painted a picture of a thriving organization: 196 members by the beginning of 1935, with another 140 candidates for membership. Many of the latter group were considered "doubtful for one reason or another," but 50 were viewed as altogether desirable: quite conceivably two of these were the world's most famous seismologist and his wife.

Neither the original group nor the camaraderie would last. In 1935 the Glasseys sold their interest in the original Fraternity Elysia to McConville and moved to Roscoe (now Sun Valley), opening a new club that became known informally as "the Ranch." There is some indication that friction between Glassey and McConville led the two men to part company, but Lura's pregnancy also played a role, prompting the young couple to move closer to modern hospital facilities. Thanks to its convenient proximity to Los Angeles and Hobart Glassey's tireless energy, the new group and the couple thrived, but not for long. Tragically, Glassey was killed at age thirty-six in an accident—a fall that broke his neck—at his home in Roscoe in April 1938, by Richter's account, "a great blow to the whole organization." Lura Glassey continued to operate the group: through the 1940s Richter's correspondence with a number of individuals refers to "the Ranch," as a place clearly near and dear to many hearts. A former member, Virginia Jurgens, wrote to Richter that, even after she had joined the military and moved to San Francisco, the "Ranch forever seem[ed] more like home than any other place."

Lura Glassey, however, found herself facing, in Richter's words, "years of increasing legal persecution, which ended in her serving a jail sentence and finally losing an appeal to the U.S. Supreme Court." By the account of Flo Nilson, the owner and informal historian of a Southern California nudist retreat, Mystic Oaks, that traces its lineage to McConville's early camps, authorities targeted the Fraternity Elysia after a member committed suicide. Dawn Hope Noel, daughter of Broadway actress Adele Blood Hope, killed herself with a .22 rifle in 1936. To many not terribly open minds of the early twentieth century the conclusion was easily reached: this young woman's dissolute ways had led to her tragic demise. The *Los Angeles Herald-Express* ran an exposé of nudism a week after Noel's death.

The case to which Richter referred was *Glassey v. State*, 1947. The facts of the case are as follows: Police officers using false names appeared at the entrance of the Fraternity Elysia, a property bounded on three sides by uninhabited hills, paid the customary visitor's fee, and signed a registration form acknowledging acceptance of nudism as a wellspring or fountainhead of moral and health benefits. Observing a number of nude men, women, and children engaged in activities such as badminton, swimming, and sunbathing, the officers arrested a man and a woman who were acting as managers and charged them with violating an ordinance proscribing operation of facilities patronized by three or more nude persons not of the same sex. (Two nude badminton players not of the same sex would have been acceptable, it seems.) A municipal court found both guilty and the judge imposed sentences of 90 and 180 days for McConville and Glassey. The Supreme Court ruled that the appellants had not demonstrated that the ordinance unduly restricted personal liberties. The original decision stood.

The case became a cause célèbre within the nudist community. In 1948 supporters of Lura, led by a woman who signed her name only as "Alicia," published an announcement in *Sunshine and Health*, the descendent of the earlier nudist magazine *The Nudist.* The announcement included a personal letter from Alicia to Lura and enjoined the nudist community to support the continuing legal battle: "*let nudists realize* that '*one of us*' is complying with all possible legal angles involved, by permitting herself to be returned for indefinite periods to the city jail in 'consenting to cooperate with the ASA and ACLU.' What courage that takes!—And in a *mother's* private life!"

In Richter's eyes this episode represented "about as nasty a little story of the failure of minority group to secure its constitutional civil rights as I know of." Richter might not have been a natural nudist, but be does appear to have been a civil libertarian, if not by nature then certainly by nurture. In 1976, for example, he sent twenty-five dollars to the Heart Fund and the Red Cross, the latter earmarked for earthquake relief in Turkey; fifty dollars to the Stanford Alumni Fund, and one hundred dollars to the American Civil Liberties Union. By 1976, he of course had more than *Glassey v. State* to fuel his libertarian leanings: he had also endured the painful experience of losing a cherished family home to the state-sponsored steamroller of progress.

Lura Glassey's legal troubles dampened but scarcely ended Richter's career as a nudist. He and Lillian visited other area groups but, while he "enjoy[ed] visiting them and meeting some of [his] old friends there," found that they had "lost something of the old quality." The new groups were, in Richter's words, "family groups of rather ordinary, wholesome people with the decent minimum of education." The individuals who had been part of Glassey's group were, it seemed, something else. Although Richter does not elaborate, one infers the members of the Glassey group to have been well-educated individuals, very likely iconoclastic in nature. Glassey himself was a psychotherapist who had studied psychology at Syracuse University in New York, although he did not earn the Ph.D. that some writers later bestowed on him. The movie *Elysia* and his articles in the *The Nudist* reveal him to have been impressively articulate in spoken as well as written communication. Glassey's partner, Pete McConville, played the role of camp gatekeeper in a later film, *The Unashamed*, directed by Allen Stuart. A 2000 review called this movie "unique among nudist exploitation films," explaining that all such films "preached the benefits of non-lewd social nudity." (All such films were considered "exploitation films.") The review goes on to say, "Issues of adultery, racism and suicide all make this a prime example of an exploitation film, but they also make the film stand out as the only nudist 'volleyball epic' to address human experience in anything like realistic terms. Amazingly, nudism still comes across as pure and life-affirming." The film is also educational: one learns that, at least in rugged outdoor settings, nudists do wear shoes. But, the "pure and life-affirming" assessment notwithstanding, the film is surprisingly dark, its young heroine is by the end of the movie driven to stand atop a mountain peak, contemplating life—more specifically, one infers, taking her own.

The Unashamed was filmed several years after the Richters joined the group. Viewing the many extras who appear in the film, one does not find any obviously familiar faces. One wonders, however, if one or both might not have been among the many extras filmed at a distance or from the back. Decorum would have probably prevented Richter or Lillian from appearing recognizably in the flesh, so to speak, but it would have been entirely in keeping with his wry sense of humor to have allowed the back of his head—or the rest of his back—to appear in a nudist volleyball epic.

Glassey's partner, Pete McConville, does appear in the film, although frontal male nudity remained at the time outside the bounds of acceptability for even an exploitation film. (Early nudist magazines were another story.) McConville originally hailed from Dublin, Ireland, and, prior to his life as a California nudist, had pursued disparate adventures around the planet, including fighting in the Boer War and running a grocery store in New York. *The Unashamed* reveals him to have been short, white-haired, and to some extent wizened (one suspects the sun will do that to a body), but still nimble enough to hop on a bicycle and pedal down dirt roads with ease. He spoke in lilting Irish English with a light accent. He continued to run nudist camps after Hobart Glassey's death. In 1954 he sold his camp, by then renamed Olympic Fields, to Wally and Flo Nilson.

In the mind's eye it is not difficult to fill out the rest of the portrait of the people who became Richter's first close friends in life: talented iconoclasts, fellow square pegs in a world full of round holes. Could Charles Richter—he of towering intellect and equally formidable demons—have found a real sense of kinship with any other sort? He continued to correspond with many of them for decades, most notably Lura, who had remarried and taken the last name of Broening. (Interestingly, while Richter's closest colleagues knew him by the seemingly affectionate nickname Charlie, his letters reveal that, among those to whom Richter truly felt close, he was always

Charles.) The letters spoke of individuals who must surely have been former members of the group; at times they extended invitations to social gatherings. The camaraderie appears to have continued in later years on an informal basis rather than under the auspices of an organized nudist group. One assumes these later gatherings were clothing-optional affairs; the invitations themselves do not, however, specify.

Colleagues who visited Richter at his home sometimes found him clad in a pair of jeans but nothing else, and were left to wonder if clothing was optional within the Richters' household when guests weren't around. (Former neighbors on Villa Zanita were, at any rate, aware that the Richters were nudists. Having moved into a house with a large wall of windows in front, it appears to have not been a particularly well kept secret.) Nor were Lillian's nudist ways any secret in family circles. Bruce Walport recalls family retreats during his childhood when Aunt Lillian sometimes sauntered stark naked into a kitchen full of people, retreating only after being told in no uncertain terms to go put some clothes on. Both Lillian and Richter apparently also practiced nudism in the backcountry. According to a brief article in *Nude and Natural* magazine, "3000 feet of films documenting the nude hikes of Richter and his wife Lillian" were destroyed when fire swept through Bruce Walport's Granada Hills home after the 1994 Northridge earthquake.

Richter might have described the Glassey group as providing his first real experience with intimate friendship, yet he had clearly found earlier kinship with his wife, who by some accounts worshipped her husband. Yet herein lies another tale that grows vastly more complex as soon as one delves beneath the surface. The *Los Angeles Times*' writers who crafted Richter's 1985 obituary interviewed a number of his former colleagues and acquaintances, many of whom offered recollections of the marriage, all of whom declined to be named. The obituary quotes a woman who had known both of the Richters: "They each got what they wanted from the marriage. Lillian got a feeling of security, of belonging, of having someone to look up to, and Charlie got someone who worshipped him. I must tell you: He was very proud of her and her writing."

Lillian Brand Richter had in fact been not only a writer but also a writing teacher for adult education classes at a number of local high schools and colleges. She taught for the adult education program at John Muir High School in Pasadena as well as other area high schools; she taught at the Maren Elwood Writing School in Hollywood. For many years one of the mainstays of her life was a creative writing night class that she taught at Glendale Junior College. One-time GJC student, Warren Boehm, recalls a teacher who had been tremendously kind and encouraging to students who were interested in publishing their own articles. He also recalls her as "feisty": like her husband, Lillian did not mince words. But also like her husband, blunt did not mean unkind. Boehm recalls Lillian's genuine interest in teaching her students not only how to write but also how to get their work published. She once said that "people are my hobby and teaching is my life's work."

Elaine Simmons, who first met Lillian in one of her classes, came to "savor her droll sense of humor and her airy-headed treatment of things she thought were not worth bothering about." At one point Lillian turned to Simmons for help finding a real estate attorney during the eminent domain proceedings. Simmons recalled how the Richters had not only sought legal advice but also thrown a block party on Kenneth Way to make sure their neighbors were aware of their rights. "If such a project needed doing," Simmons, herself very much a can-do woman, wrote in a 1985 letter to the *Pasadena Star News*, "you could turn Lillian loose on it, and know that it would go full steam ahead."

Lillian's own writing was published in a number of magazines, including some articles written under a male pseudonym for men's publications such as *True*. Lillian once bet a friend that she could write and sell an article on any topic the friend might suggest, which turned out to be salt. She did indeed write and sell the piece to a children's magazine, although she later joked that she made more on the bet than she did on the article. By her niece's later account, while Lillian aspired to fame as a writer, she also remarked that she had made a career out of "collecting rejection slips."

Writing was a shared passion for Lillian and Charles. Yet for all of their kinship, and the supposed worship on her part, it was clearly an unusual union. Both Bruce Walport and his sister, Dorothy Crouse, recalled a couple who frequently went their separate ways. The couple often spent their Christmases apart, Charlie spending his free time hiking alone in the mountains while Lillian flew off to some exotic locale. In summer of 1965 she arranged a trip to Timbuktu, "just because she had become obsessed with the name of that city in Mali." According to Walport, Lillian had been the first white woman to travel unescorted to Timbuktu. She and Charles exchanged almost daily letters during this trip, which began with a week in Paris before she left for Africa. In Paris she enjoyed the art galleries and the local art studios but had no use for the many antique shops, whose wares she described as "pure junk." ("You know how I hate antiques," she wrote in one letter.)

By June she had traveled to Dakar in the Republic of Sengal and Bamako in Mali. By June 4 she made it to Timbuktu, which she described as "exactly as it was thousands of years ago." Her letters describe a whirlwind of exotic sights and adventures, all devoured with zeal: stunning Senegalese women with snow white cloth draped over black skin; the natural beauty of Dakar, her first view of the Southern Cross in the sky over Timbuktu, its brilliance unmarred by any nighttime glow from the earth. In one letter to Richter she observed that her French was better in Africa than it had been in Paris: he responded that it was one of her "priceless Lillianisms," and that he understood exactly what she meant.

That Richter did not join her on her global adventures, and that she did not join him on his mountain hikes, suggests a certain distance in the relationship. There was more to this story than what outsiders saw, but we will get to that part of the story later.

Charlie and Lillian were, it seems, well matched in their awkwardness. A former Caltech colleague related that he had attended dinner gatherings at the Richter home on a couple of occasions, "but they were terribly awkward evenings." They were well-matched in other respects as well. Dorothy Crouse recalls an Aunt and Uncle who, "reminded [her] very much of each other."

An individual in her own right far more than the traditional woman behind the great man, Lillian never involved herself with Richter's professional life. Even his closest colleagues knew her only in passing, and recall little of her interests or her life. Richter's former colleagues often describe her with words like "eccentric." Hertha Gutenberg called her "all right, but a little peculiar," and added that she wrote for a tawdry romance magazine. A few of Richter's former colleagues hint that Lillian had not been considered a properly supportive spouse. If Richter walked to the lab, or if Lillian dropped him off and kept their one car herself during the day, they came to their own conclusions.

Betty Shor, who had been a colleague as well as a grad-student wife, provides a different perspective. Faculty wives had well-defined roles in the middle of the twentieth century, especially at schools such as Caltech, where the tenor of academic life tended to be especially formal. Apart from the expectation that they would as a matter of course provide a bedrock of support for their husband's careers, wives were expected to be active in women's groups that could involve a great deal of social climbing. In Shor's words, Lillian was "just not interested." She had her own interests, her own profession, her own friends.

In 1979 Richter himself recalled the early days, when "a great deal was very formal in the way of dinners, etcetera. It was customary at the Institute that one had to put on full dress so that the ladies could show off their evening gowns. This was very unpleasant and contrary to my feelings, so I was relieved when it finally petered out." Pressed by the interviewer, "Maybe your wife enjoyed it though?" Richter replied, "Not particularly." Although Dorothy Crouse recalls her aunt as having been well dressed, the formal party scene at Caltech was not her cup of tea.

Shor also recalled the story her father had passed along one evening when she and George had dinner at her parents' house. Her father had "commented that there had been a geology faculty meeting that day at noon. For lunch Richter pulled out an egg, tapped it heavily against the counter for peeling—and a raw egg slid down the table to a halt in front of the stern department chair, John P. Buwalda. Richter muttered, 'I thought it was hard-boiled when I got it out of the refrigerator.'" The story captures the essence of Richter's well-intentioned but often hapless bumbling, but it captures something else as well. "The implication," Shor adds, "was that he had packed his own lunch, which the others found amazing."

Hertha Gutenberg's account can also be considered afresh, with Shor's perspective in mind. The consummate faculty wife of the consummate academic scientist, Hertha would have undoubtedly been among the reigning aristocracy of her social circle—the circle for which Lillian had no use. When Hertha recorded for posterity the observation that Lillian wrote for "*True Romance* or one of those magazines"— neglecting to even mention Lillian's long career in teaching—she might not have known the difference between the women's romance magazine and the men's magazine *True*, for which Lillian wrote under a pseudonym. Although its articles sometimes ran towards the outlandish, *True* also published serious articles about science, nature, and current events.

Hertha may never have read the delightful essay that Lillian published under her own name in *Recreation* magazine in 1949: "Writing is Fun." The article itself belies the lightness of its title, opening with the paragraphs,

> Whether you ever sell anything you have written or not, it is fun to create a short story. In real life you cannot influence people's actions very much, and there is nothing whatever you can do about Uncle Ben or the woman next door. But on paper you can do just as you please with your characters! In fact, if Uncle Ben is worse than just a harmless eccentric, you can have him murdered for his vile temper, his meanness with money, his slandering tongue.

> Now, who could have murdered Uncle Ben, and why? Give your murderer strong motives, and a good brain with which to cover up his tracks, and go on with your story.

The short but well-crafted article goes on to point out that any real-life expertise could be parlayed into publishable articles. It then described in some detail the process of finding appropriate publication venues and sending query letters. It closed with the words, "While you are struggling to get Uncle Dick onto paper you'll be too busy to worry about unpaid bills, or to be bored with life. What if editors and publishers never appreciate your writing? You have had fun doing it, haven't you?" Lillian might have struggled to establish herself as a serious writer in the middle of the twentieth century, yet clearly this was a woman—a writer—with interesting things to say.

Attempting to piece together Lillian's career as a writer, one's thoughts again hearken back to Virginia Woolf. The challenges of building a career—of building a *name*—as a writer, daunting under the best of circumstances, were very much amplified for women of Lillian's era. For some publications she could not use her own name at all. Her own name had moreover evolved over the years, from Brand to Saunders and back to Brand, and finally to Richter. While she continued to use her maiden name professionally during her marriage to Richter, it is possible she used her first married name for some of her earliest published works. A writer by name of Lilian Saunders published a series of poems in the literary magazine *Poet Lore* in 1925 and 1926–a time when Lillian Brand would have been Lillian Saunders. A half-dozen poems published in 1925 involve similarly romantic themes, all woven around similar imagery: the sea, long gray waves, rushing breakers, phantom ships. For example,

> SEA JEWELS
>
> Blue of the turquoise, luminous and blue
> As turquoise would be if the sun shone through.
> Green of the jade, where great waves curl and make
> A glistening cavern just before they break.
> Foaming with pearls the white spray dashes high
> Snatching a shower of sapphire from the sky,
> And over all gleam diamond sparkles shed
> In glorious largesse from the sun o'er head.

Poems published in 1925 would have been penned at a time when Lillian Brand was living in the Bay Area town of Alameda, a newlywed with a baby on the way, or soon to be on the way. The writer's state of mind—whoever she was—shines through the words with clarity: "'tis the sea's magic waves those scattered beams, / To make for us a path, out to the Port of Dreams."

A single poignant published poem from September 1926, however, suggests a path gone awry:

Seven O'Clock

A raucous shriek tears through the slumbering air . . .
The monster wakes!
His open jaws stretch wide
To wait the coming of the daily sacrifice
The tale of youths and maidens who each morn
Are cast into his ever greedy maw.
All day he feeds upon them, drains
Their life blood and their strength,
Sucks their vitality, gnaws at their soul,
Until when evening comes he spews them out
Pale husks, dry shreds, fit only to be flung
On life's scrap heap, to wither there, or rot.

Here again one finds precisely the sentiments one would expect from Lillian Brand in 1926, a young woman whose marriage must have been falling apart, and with it her life circumstances. Her scrap heap would be her native Los Angeles, where she landed rather ignominiously: living with the Podolsky family, "helping them out." But were Lilian Saunders and Lillian Brand one and the same? The disparate spelling of the first name raises a red flag, but young women—aspiring young writers in particular—have been known to experiment with such things as spelling of first names. Another bit of evidence, apart from the nature of the poems, suggests that Lilian Saunders was in fact Lillian Brand: the name "Lilian Saunders" disappears from the pages of *Poet Lore*—and apparently from the planet—at just the time when Lillian Brand resumed use of her maiden name.

The records of *Poet Lore* provide no help: while the publication still exists, its operation changed hands several times through the twentieth century, and records from the 1920s no longer exist. Had Lilian Saunders continued to build a writing career under that name, we would have a trail to follow. Instead, assuming that Lilian was Lillian, the trail was severed, and Lillian's burgeoning career as a writer was, if not itself severed, then certainly interrupted.

Clearly, in any case, Lillian Brand was a woman with an identity—if not an established name—in her own right. Relatives recall Lillian as having been proud of her status as the wife of a famous scientist; not above making this status known if, for example, she thought it might improve the quality of service in a restaurant. In the classroom, however, she stood on her own. Former student Warren Boehm recalled that Lillian had not even mentioned her husband's name until midway through the semester. Until that time the class had no idea they were being taught by the wife of a famous scientist. Richter's colleagues, meanwhile, apparently had no idea that Lillian was a woman of substance in her own right. She had in fact placed herself in elite company early in life. At the time that she graduated from Berkeley fewer than 20 percent of American teenagers graduated from high school and only a scant 2 percent of young adult women had graduated from college. (Young men earned college degrees at a somewhat higher rate, although still in numbers far below today's figures: in 1920 about 4 percent of twenty-three-year-old men had earned a college degree.)

Laurie Walport, Bruce Walport's eldest child, recalls Lillian with admiration, a woman who was well ahead of her time in her independence, interests, and direct manner. Lillian assumed the role of cultural director for the family, taking Walport and her young siblings on outings to the nearby Norton Simon art museum and gardens. Laurie Walport describes Lillian as "larger than life," and adding a great deal of color to the fabric of family life. Great niece Kathy Haag echoes these accounts, adding that Lillian sometimes took her young relatives shopping for nice dresses. (She does admit that Lillian's young relatives could sometimes drive their Uncle Charles to distraction with their antics, for example by jumping up and down to make the living room seismometer move. Even as an adult, children often left Richter bemused.) Laurie Walport's mother, Mary White, who was especially close to both Charles and Lillian, admired Lillian's wide-ranging interests and her degree of independence. Whereas many of Richter's close "writer friends" were women, Lillian enjoyed dancing, and danced for years with a male partner (one cannot begin to imagine Charles Richter on the dance floor). At the time such things were simply *not done*, yet White marveled at how comfortably these friendships were accepted, without as much as a trace of jealousy.

By living her own life, Lillian refused to live vicariously through her husband's ambitions and accomplishments. Unlike most faculty wives of her day, Lillian's ambitions were her own. For this, and for eschewing the social circles to which faculty wives of the mid—twentieth century were supposed to belong, Lillian was regarded—to the extent that she was regarded at all—as an unsupportive spouse, or worse. Richter's own gruff manner, which extended towards Lillian as well, did little to allay these views.

One of Richter's poems, penned in the late 1960s, hints further at the extent to which Lillian had her own life and interests. The poem, written for a woman who was not Lillian, begins:

> I need a name for you. The one you sign
> Is good, sounds well, I could not love you more
> Without it; none the less it is not mine.
> We need a private name unheard before.
> What then? A poet might be Lesbia—no,
> Not Sappho, not that bitter love again!
> My very dear, I would not have you so;
> I want a woman with a taste for men.

As we will see, Richter's love interests had grown complicated by the 1960s. Of relevance for the present discussion, however, is the seemingly inescapable conclusion that one of the women in Richter's life was more partial to women than to him. Some might consider it posthumous slander, or at least speculation, to suggest it was Lillian, but if not her, one has to wonder, who?

If Richter's poem does refer to Lillian, any number of puzzle pieces fall into place: the separate vacations, his outside love interests, their easy acceptance of each other's close friendships with members of the opposite sex, their own abiding and devoted but not terribly romantic union. The love that dare not speak its name has in recent years, as some have observed (one likes to think good-humoredly), become the love that won't shut up. But in the 1960s some loves remained not only unimaginable but unimagined. That Lillian frequently left on weekend trips with a girlfriend—the Bermuda Inn in Lancaster having been a favorite destination in the 1960s—would have raised no eyebrows: just two old friends enjoying each other's company for the weekend. That she and Marguerite Barsot—who lived in Lancaster—had been "like sisters" their entire lives would have been taken similarly at face value. Friendship, companionship, sisterly affections between two respectable ladies. What else could it have been? In retrospect one can't help but wonder. It would explain a lot.

Fig. 11.8. Old postcard of Bermuda Inn, "Lancaster's newest and finest motel."

Further clues can perhaps be found in the record of a chance encounter that two young travelers, Sonia Rosenberg and Gladys Broderson, had with the Richters. In October 1937, Sonia and Gladys took a mule tour to the floor of the Grand Canyon with a guide, "Mack," and another couple who had signed up for the same tour, Charles and Lillian Richter of Pasadena. Sonia and Gladys were struck throughout the trip by Charles, who seemed so peculiar that the women struggled to make it "through dinner without disgracing ourselves by laughing at him." Gladys, who kept a detailed journal of the trip, wrote, "Lillian confessed during one of our conversations that she didn't like he-men and Sonia was just about to say that she couldn't stand any other kind when Lillian announced that she liked the feminine-type (that's Charles)." Sonia and Gladys later decided that Lillian must have "married Charles as a noble experiment, or as the subject for her next best seller." One notes further that, if it would be fair to regard Richter as a "feminine type," the same appears to be true of Lillian's first husband as well.

Details such as sexual orientation aside, one further intriguing and less speculative insight into the Richters' marriage lies hidden within the mountains of his writings and correspondence—in particular his trip reports, the excruciatingly detailed journals that he kept during annual hiking trips to the Sierra Nevada. He wrote of a trip that he and Lillian took

Fig. 11.9. Charles and Lillian Richter on a mule tour of the Grand Canyon in 1937, with two tourists and their guide. (Photo from scrapbook compiled by Gladys Broderson.)

together in 1950 that did not go well. After a few days of hiking Lillian found herself tired, crying, and unsteady on her feet. Eventually she landed in a local hospital where a doctor prescribed salt tablets and sleeping pills.

Later journals describe solo backpacking trips. Richter would leave Lillian at home, frequently in the company of her niece's daughter Nancy Jean, as well as the household kitty of the day (in 1958, Temblor; in 1964, Shomyo). During these trips Lillian and Charles exchanged not only phone calls but also letters to and from the Sequoia campground that Richter often used as his base camp. Many of Lillian's notes were on postcards, brief and breezy messages: "The Pie and Burger place has three kinds of delicious fresh fruit pies and I've had a few pieces. Today I heard the Messiah. Beautiful! Shomyo is looking for his dinner. Love, Lillian."

One cannot, after all, write at length on a postcard; nor is one inclined to say much of a personal nature. A rare folded note card, sent within an envelope and dated July 24, 1964, reveals more. "I feel 'homesick' for [the scenery]," Lillian wrote. "Actually I think I can walk better than I used to—think I could walk to Bear Paw. But the battle with colitis is even worse now than it used to be—am not sure I could manage it there. The colitis makes all traveling so difficult for me, and expensive. Damn colitis!!!!" She went on to lament her inability to lead the healthy lifestyle that she thought would help: "I need to cut down on my 'many medicines,"' she wrote. "This worries me."

And here, suddenly, we arrive at a snippet of information that changes everything. Colitis is an acute or chronic inflammation of the membrane line of the large bowel. Doctors now recognize distinct types of colitis, including infections, caused by bacteria or other "bugs," as well as two major forms of chronic colitis, Crohn's disease and ulcerative colitis. The latter two are similar in some but not all ways: ulcerative colitis causes inflammation only in the colon and possibly rectum, while Crohn's can cause inflammation in the small intestine as well, occasionally extending to the stomach, mouth, and esophagus. Chronic colitis can take a serious toll on the body: the condition is associated with a significantly increased risk of colon cancer.

Lillian's letter indicates that her condition was chronic rather than transitory. She would have suffered for years with symptoms of her condition: abdominal pain, diarrhea or constipation, painful spasms, lack of appetite, fever, and fatigue. Over time the condition causes progressively deeper and larger ulcers within the bowel, sometimes even puncturing the bowel walls. The cause of both ulcerative colitis and Crohn's is unknown, but may be caused by an immune system run amok. In particular, the immune system may become activated in the absence of any outside invader such as harmful bacteria or viruses, essentially turning the body on itself via chronic inflammation. Inflammation is a key element of the body's natural defense against trauma: for example, the swelling that naturally immobilizes an injured joint.

Colitis can be aggravated by diet and controlled to some extent by avoiding particular offending foods. Even today, however, effective treatments are few and far between. Certain compounds can act topically (via direct contact) to reduce inflammation; oral corticosteroids, or "steroids," can reduce inflammation throughout the body, and antibiotics can reduce inflammation. None of these treatments is, however, reliably effective and safe for long-term use. Long-term use of steroids in particular has predictable and deleterious side effects.

Colitis is, simply put, miserable. In addition to the physical discomfort, persistent diarrhea can present an enormous handicap in social settings, and, one imagines, a virtually insurmountable difficulty in the wilderness.

Other colitis sufferers—including, by her niece's account, Lillian—experience constipation so severe that bowel movements are not possible without enemas. Colitis sufferers are prone to dehydration and fatigue or weakness associated with chronically poor absorption of nutrients. One understands why, as Richter wrote in 1949, Lillian had "lost weight and become physically less attractive." The photograph from the Grand Canyon suggests that this might have been around or before 1937: at least in this one photo, Lillian appears gaunt. According to the journal of Gladys Broderson, their guide Mack had observed after the Richters left the group, "Ah knowed there was somethin' wrong with her when ah first saw her, her nose is too short." As Gladys observed, it wasn't clear what the length of one's nose has to do with anything, but perhaps the guide picked up on other clues, hinting at a more general lack of well-being.

People typically do not discuss a condition like colitis in either polite or even fairly intimate social circles. Had Lillian not written a small number of especially honest letters—had Charles not kept them—one would be left to interpret Lillian's absences on his summer hiking trips in an altogether different light. From this one fact a different understanding emerges: she was not necessarily avoiding him, she was avoiding an activity that she very much loved but could no longer handle physically.

The onset of chronic colitis varies considerably between individuals. The disease afflicts an unfortunate small percentage of sufferers in early childhood: the condition can at its worst be life-threatening. The condition is more common in women than in men, and the most common age for the condition to appear is between fifteen and twenty-five years—and nearly always by age forty. It is therefore almost certain that colitis contributed to the bad days that Lillian experienced on her hiking trip with Charles in 1950, and that by this time she had been suffering with the condition for many years. Her letters from the mid-1960s reveal that she continued to suffer late in life. Her travels in France and Africa were, she wrote, complicated by the need to "manage complex bathroom arrangements." A letter from 1965 reveals something else: that she had passed her unfortunate genetic heritage to her son. "If my son had inherited a decent constitution," she wrote, "he would still be with us. No one commits suicide who feels full of vim, vigor, and vitality—no matter what happens!"

Lillian was quite possibly sick during, or even before, early adulthood, which could help explain an earlier mystery. It is far less unimaginable that a young woman in the early twentieth century would not raise her son from her first marriage if that woman were seriously and chronically ill.

Colitis sufferers can experience periods of remission, either partial or total: it's possible that her early years were even worse than the later ones.

By her niece's account Lillian suffered other serious health problems as well, including "female problems" severe enough to prevent her from having a child with Richter. By Crouse's account, she and Richter would have liked to have had children, but were unable to.

Lillian also suffered a less serious but irksome medical condition: a serious allergy to poison oak, an unamusing biological specimen that abounds in the hills of California, including the then-remote part of Altadena where the Richters moved in 1969. Their cat Shomyo—described by Elaine Simmons as a "huge, white Persian"—went outside during the day to stalk small creatures that lived in a nearby dell. The cat would then come inside and settle comfortably on her mistress's lap, and Lillian would later break out in a viciously itchy rash from her fingertips to her elbows.

Poison oak is a tiresome but transitory inconvenience; Lillian's chronic health problems were life-altering. The simple fact that she suffered from colitis, as well as other more typical female health problems, explains a great deal about a relationship that was by all accounts complex, but perhaps in some respects not quite as persistently strained as it appeared. One of Richter's letters to Lillian during her 1965 trip sheds further illumination on their separate travels: he wrote about his growing distaste for and inability to cope with the vicissitudes of air travel.

Still, Richter's 1949 letter to Doctor Moriarty does reveal substantial strain between husband and wife. This letter leads to another chapter of Richter's life: one of the more complicated chapters of his enormously complicated life story; one both apart from and germane to the story of his enigmatic relationship with Lillian. And so we move now from the story of the love of Charles Richter's life to the rather more complicated story of the loves of his life.

Richter's Women

Only a few have loved me, but those few
Are no more like by night than by day
—*Charles Richter, "For Julia," 1966*

The final page of Richter's 1949 letter to Dr. Moriarty, typed on a different typewriter and apparently added only after explicit prodding, tackles matters related to sex. "Part of my sex behaviour," he wrote, "is governed by a fear—the fear to face a situation which might conceivably exist, but which I cannot believe is true to any important extent. It might be so that I married Lillian not because I cared for her, but because I needed a woman and she wanted me; that I never really loved her; that eventually, ten years later, I fell genuinely in love with Margaret, who does not seriously care for me."

Full stop. Rewind tape.

Margaret. Only a single person by that name appears previously in the letter, and that Margaret we have already met. It seems grossly at odds with all of his other expressions of proper brotherly feelings—concern, duty, occasional annoyance—to suppose that he felt a strong degree of romantic love towards his own sister. Another Margaret does make an appearance in Richter's papers: Margaret Murphy, a member of the Glassey group. Yet earlier in the letter to his doctor he refers to Margaret, explicitly noting her to be his sister, which seems to leave little room for doubt to whom the later part of the letter refers. One would think that, were the second Margaret different from the first, Richter would have been careful to note this.

So we face the question: could he have fallen in love with his own sister? The phenomenon is known to exist. As Glenda Hudson argues in *Sibling Love and Incest in Jane Austen's Fiction,* "The joint experiences of shared childhood and mutual associations . . . create a potent and sympathetic love, a co-mingling of fraternal and erotic feelings, which, although the emphasis is very much on the former, we must recognize as a kind of incestuous love." Certainly Charles' and Margaret's shared childhood was far more tightly knit, far more insular, and far more peculiar, than most. He had referred to her in his 1927 journal as "my sort of woman." And perhaps tellingly, even for a man who shared little of his personal life with his colleagues, Richter remained almost completely mute on the subject of his sister. Richter enlisted the help of a business partner to clear Margaret's belongings out of her apartment following her death in 1979; he had not previously known of her existence. And for all of the time that Bruce Walport spent at the Richters' home, he recalled having met Margaret once or twice. "The topic didn't come up," he said.

Recall also that Richter had essentially followed Margaret to Stanford. One finds oneself hard-pressed to interpret Richter's words in any way other than how he wrote them, which would indicate that his affairs of the heart were complicated indeed. Richter's 1949 letter indicates that he fell in love with Margaret ten years after his marriage to Lillian, around 1937–not long after Margaret moved back home from Arkansas. In the 1949 letter Richter went on to say, "There are times when I wish I could be in love with Lillian instead of Margaret." The attachment was, it seems, no passing fancy.

But which Margaret? And were the feelings reciprocal? One can turn to Margaret Richter's poems for possible clues. One from 1937 suggests that she was in love with somebody:

CYPRESS

Love has come tonight like the winds of ocean,
Striking dark blue waters to foaming whiteness,
Raising great green breakers to sweep the sea
Shore,
Driving the fog-banks;

Twisting cypress trees into gnarled, fantastic
Monsters borne to earth in an ancient struggle,
Striving nobly, waving their torn limbs fiercely,
Threatening vengeance—
Winds that send the fog through the shattered
Forest,
Hiding havoc wrought by their strength, and bringing
Riven branches, seen in the fitful moonlight,
Garments of beauty

In love with somebody, but who? We do not know, and probably we never can know. One wonders if one really wants to know. A much later poem provides possible further clues, however, in its words as well as its presentation:

Bright as the Star

Let me bewitch you for a season
And make you merry to unreason;
Fruits and feasts and reveling,
Pine cones for eternity;
Bright as the berries on the bough,
Bright as the jewels on Circe's brow,
Bright as their presiding star
Be these last days on the calendar;
Life be a gilded masque,
Bacchantic task,
On earth our stage—
Turn the page!

The poem was typed on a small white card with no date or other information. Judging from the papers that Richter apparently filed around the same time, the poem was written in the mid-1960s. Turning the page one finds a handwritten inscription, "Charles, from Margaret." One can perhaps debate the meaning of the poem, but its tone does not impress the reader as entirely sisterly. In the complex and enigmatic story of Charles Richter's life, nothing is quite so enigmatic as the depths of the bond between him and his sister—on the one hand unthinkable, on the other, perhaps not. (And if one dares to consider it thinkable, one arrives at another possible interpretation of whom Richter referred to when he wrote about "that bitter love." Richter's papers include none of Margaret's personal correspondence, but one thing is clear: she never married.)

We know other things as well. Margaret left Los Angeles for Columbia University not too long after the time that Richter fell "genuinely in love" with a woman by that name. Could the timing of latter occurrence have inspired the timing of her move? By 1949 Richter noted to Dr. Moriarty that Margaret did not "seriously care for" him. As we will see, there is compelling evidence of his attachment to other women, including one serious relationship in the early 1940s. Richter does not mention any other relationships in his 1949 letter, however, leaving the reader to wonder—to put it mildly—why a man would confess to an incestuous relationship but hide a "merely" adulterous one. (The wording of the letter does suggest that the doctor might have already been aware about the relationship with Margaret.)

Richter goes on in his letter to tear down his own straw man regarding his wife, or at least to make the attempt. "None of these points are completely true," he adds after his confession. "There might have been—probably was—an interval when I did not really care much for Lillian, but went on with her largely from force of habit, and from unwillingness to make the effort to break loose. It is true that when she lost weight and became physically less attractive, and emotionally more unpleasant and demanding, that I felt very little interest in her, and considered that I was remaining with her largely because of the feeling of social criticism that might be directed at me if I were to leave her under such circumstances." Honest sentiments, if not especially admirable ones. He goes on to say, however, that "by remaining with her through all this, I have ended by being quite genuinely attached to her. Many of the deeper and worthwhile elements in Lillian's character, which formerly I could only guess at, have come up to or near the surface, and she is much more of a real person than she was." One cannot help but wonder if Lillian had become more of a real person, or if Charles had simply grown to a point where he could see and appreciate the person she had always been.

The letter closes with the observation, "It is true that when I am with Lillian my imagination often wanders to other loves, and particularly to Margaret; but that is not always the case, and there is a real response to Lillian herself, particularly when she is most herself." Here again the tape player comes to a stop once again: "other loves"? But first things first.

Affairs of the heart are complex, but romantic feelings about siblings remain well outside the bounds of commonplace complexity. Suggestions of incestuous relationships leave a person uneasy if not queasy: how could one think such a thing? How could one say it? Could another, less sordid, explanation account for his letter and her poems? Certainly. And yet of all that Richter wrote about the women in his life, few statements are less apparently ambiguous than those that he wrote to Dr. Moriarty about Margaret. Her own poetry seems to provide at least a measure of corroboration. The inference may leave us squirming and may in fact be wrong, yet the evidence seems to point in that direction.

Reading Richter's papers, however, one does have to wonder: could all of the "evidence" have been his little joke on posterity? This was a man who kept his diaries, letters, and poems, and moreover made a decision to donate his papers—all of them—to an archive where they will remain in safekeeping forever. Yet time and time again one arrives at vexingly inscrutable ambiguity. If the obvious interpretation is in fact the wrong interpretation, one is inclined to wonder if maybe the joke is on us.

Ample evidence points in other, more typically sordid, directions. Richter's papers do not include correspondence with either his sister or Margaret Murphy, but do include stacks of other letters sent to and from other women. A 1941 letter addressed, "Kim darling," sheds light one of the other serious relationships in Richter's life. "This will be the letter," he wrote, "which I have been menacing you with for a long time. I suspect it will undergo rewriting before it gets sent to you. . . . I hardly need to insist that you must destroy this after reading; it would be poison for both of us if anyone else got their hands on it." (The biographer notes that Richter did not follow his own advice.) He went on to suggest that, if she wanted to preserve parts of it, she could perhaps incorporate it into a fictional composition. Thus do we learn that "Kim" was a fellow writer.

Richter continues, "Before anything else—I love you, and will always love you. That is quite different from saying that I am physically in love with you; which, so far as I can judge of my feelings, is no longer the case." He tells her he has "emerged from the worst of it," that he has to accept that her writing keeps them apart. And yet the feelings remain: he then talks about a fear that, should they see each other, they "might start the whole thing over again." Richter observes that this would be difficult for "Bill," for whom Kim apparently cares deeply.

"You know," he writes further, with stunning if not brutal honesty, "that I manage to see and be companionable with other women I have loved—but that is mainly due to the fact that we have been concerned with other persons since." (The question pops to mind: how many women were "the rest"?) He goes on to add that nobody has taken her place in his life, that "I was a long time finding you, and you are not the sort of person form whom a substitute can be found on any street corner."

The letter continues::

> You should realize that there are such interludes in most normal lives—episodes that for a time open possibilities outside the usual limits, that present one with alternatives to one's settled ways, and that seem like a dream on [sic] an aberration when they are remembered. . . .
>
> You know, of course, that I was very close to offering to give up everything and marry you. I am glad now that it did not go so far—more for your sake than mine. If it had wrecked me—well, no great loss. But it is pretty clear now that we could have gained nothing; the war would have separated us, and we should both have made sacrifices to no immediate purpose.

He goes on to suggest that his failure to commit to her resulted from the fact that she had never "definitively broken away and come to me." Had she done so, he suggests, "The desire to see that you were well taken care of might even have supplied me with the stimulus to persistent work and seriously accomplishment which I have thus far lacked." The letter is undated, but he refers to himself as forty-two years old to her twenty-three: he wrote this sentence years after the seminal scientific contribution for which he is now remembered. The next bit of the letter can only leave one wondering. "My picture," he wrote, "of human relationships has never been restricted to those involving only two people; I have seen too many exceptions—Forgive me for bringing all this up; it is over, but I didn't want you to think I was so stupid as not to see the implications." The reader eavesdropping across the years can only wonder: What implications? And what kind of relationships was he talking about?

One can also only wonder about a line late in the letter, in a section that seems to have been added later and on a different typewriter. In this part of the letter Richter reflects on her future relationships: what she might have gotten from their relationship, and what he got from it himself. "You educated me," he wrote. The line that leaves a reader wondering (all over again) comes towards the end of this part of the letter: "There may some time be a person who will take the place in relation to me—and Lillian that you did . . . it seems unlikely." From this one is inclined to suspect that Kim's relationship with Lillian was significant in its own right; although here yet again, we can only guess at the details.

The third page of the letter ends on a different note from what one expects. He includes a love poem that ends with the lines,:

> Yet it needs resolution and cool strength,
> Here in the warmth of evening as we ride,
> To pass all comprehension, and at length
> Humbly to take love as our only guide.

An awkward poem to be sure, but perhaps a sweet sentiment in the end. Then, following a handwritten date of August 29, 1941, he adds the closing paragraphs:

> It has taken me hours to turn this out, an [*sic*] I am not satisfied with it; not even sure that I agree with what it seems to say.
>
> It doesn't seem possible that I saw you last only this morning. I hope it won't be long until I see you again; it will seem long, anyway.

> In case you want to get in touch with me, my schedule for the weekend is (at present) as follows:
>
> Saturday-morning, at the laboratory.
> Afternoon, first home, and then to the ranch.
> Evening, and all Sunday, at the ranch.
> (Lillian will be out of town Sat. and Sun.)
>
> Will stay over at the ranch Sat. and Sun. nights, but probably have to go to Pasadena Mon. to pick up Lillian . . .
>
> Much love.

And thus the remarkable letter ends not as it began, as a "Dear Jane" letter reflecting on a once-intense relationship now safely in the past. Who was Kim, one can't help but wonder?

She and Richter continued to correspond regularly for several years. She wrote about Bill, to whom she was initially attached but not married. Her letters describe how she made up her mind to leave her free-spirited youth behind and settle down with him: they were married in June 1942. In August 1942 she asked Richter if he had been to "the Ranch" lately. She also wrote that she considered it a "closed chapter in her life," and one would stay that way: "I associate it with myself the way I was a year ago—and that self I hate." She wrote, "In a way, you saved my life with Bill for me—listened to my ramblings and calmed me down and fed my starved sentimentalism!"

Kim wrote of her growing attachment to Bill and her appreciation for him as a kind and caring husband. Little by little, she began to pull away. In April 1945 she wrote to Richter from Baltimore, where she and Bill had moved. Bill had opened a package that Richter had sent and, as Kim wrote to Richter, wanted to know "1) who you are, 2) why the hell you should suddenly remember my birthday, and 3) what the hell you meant by 'May I?', etc." She told Richter that she would be happier if they did not correspond further; at that time it appears the chapter finally closed for good.

For all of this correspondence we are left knowing little about Kim, whose last name never appears on her correspondence or his. We know only that she was born in 1917 or 1918, married a man named Bill in 1942, moved to Baltimore by 1945, was a writer in her own right—of fiction it seems, although perhaps not only that—and was forced by either Bill's commitments or her own to leave the Pasadena area during World War II. She had also been a member of the Glassey group, although perhaps only briefly. Beyond this we know little except this: she was one of the loves of Richter's life.

As the correspondence with Kim petered out, other correspondences emerged to fill the void in Richter's life. In 1944–45 he wrote many letters to a woman named Mavis, another member of the Glassey group. Mavis impressed Richter as having a keen aptitude for mathematics: his letters include both mathematical discussions and encouragement for her to pursue her studies. They also wrote at length about more philosophical—more intimate—matters. In a long letter dated September 14, 1944, Richter shared his attempts to come to grips with his station in life: "Middle age," he wrote, "is the point at which one begins to be more interested in learning to use the stock [of experience] one already has. That use, in middle age, is a forward looking one; the background of the past is drawn upon to enrich the present and prepare it to serve in its turn as a past background." Later in the letter he wrote, "There is less hurry and haste than ever about me now. Everything that comes to me is so well provided with possibilities for creative experience, that I find no occasion for impatience and anxiety. Living is too full now for me to be selfishly reaching out for more. But neither is there any shrinking from what experience does come." The letter ends on a far less introspective note, with Richter elaborating on his view that "to say that you are no ordinary person is merely to state the obvious."

The beginning of this letter reveals one—but clearly only one—of the ways that Mavis was extraordinary: Richter confessed that he found her distractingly attractive. While nothing in their correspondence points unambiguously to a physical affair, one does have to wonder if she is the beautiful young woman with shoulder-length blonde hair whose nude photographs Richter kept in a black folder from Murrillo Studios in Los Angeles. Or perhaps this was Kim. At any rate, Richter's correspondence reveals emotional if not physical intimacy with Mavis.

A journal entry from September 1938 suggests that Richter's midlife indiscretions might not have been considered indiscrete by Richter himself, or by Lillian. "Now," he wrote, "I have, jointly with Lillian, to further develop a relationship which we set up in the face of criticism, and to carry it on in our way; not someone else's way." He does not elaborate on what he means by "our way": clearly possibilities run the gamut. Among these possibilities, however, is an obvious one. "*There may some time be a person who will take the place in relation to me—and Lillian that you did . . . it seems unlikely.*" Had Kim somehow been the catalyst for the transformation of an unhappy conventional marriage into a happier unconventional union? Richter's words suggest that he and Lillian came to certain arrangements around 1937, although they do not spell out the details of the alchemy.

Later in Richter's life, in his early to mid-sixties, he wrote a series of poems to a woman he called Nerissa, as well as one rather astonishingly crass expression of feelings towards another named Julia. These we discuss in the following chapter. Of relevance to the present discussion is the inference that Richter had a relationship with a woman whom he called Nerissa and yearned to have one (although it might never have been consummated) with Julia. He also appears to have been involved with a third woman before and after Lillian's death. The bond with Margaret (again, though, the question, *which Margaret?)* appears to have endured for decades as well, at least intermittently.

One is left with few real clues to the identities of the other women in Richter's life: only a few first names, most of which are common and all of which, as we will see, might not have been the women's true names. It is moreover, unclear, how many other women there were.

One letter in Richter's files leaves the biographer wondering if he was a hopeless flirt or simply hopeless. In a letter dated August 15, 1969, he thanks the recipient of the letter, Miss Boyer, for sending a copy of *Los Angeles* magazine with her story about the Seismo Lab. In the letter Richter admits to not caring for the headline, a play on the phrase *fun house.* In pedantic fashion he goes on to explain, "In my recollection, a fun house is an establishment such as used to be found on piers at beach resorts." The hopeless part, however, comes at the end: "I enjoyed talking with you, not to mention looking at you." One must, of course, keep in mind that the year was 1969: simpler, vastly less politically correct times. Susan Newman, the longtime executive director of the Seismological Society of America, recalls Richter's flirtatious behavior towards her and her young female assistant in 1976, when Richter received the Medal of the Society. Newman emphasizes how different the times were: her professional stature notwithstanding, in gatherings of the almost entirely male membership of the society, she not uncommonly had her butt pinched. The closing of Richter's letter, and his flirtations (as an esteemed and decidedly senior scientist), clumsy as they may have been, were probably harmless.

There is considerable evidence that, his complicated affairs of the heart aside, Richter was ahead of his time in his ability to work shoulder-to-shoulder with women in a professional capacity, and without hint of impropriety: Vi Taylor, Gertrude Killeen, Betty Shor, Karen McNally. According to Richter, Vi Taylor had been "scared pink" when she first took the job, but proved to be most capable, and "the place was never run so well. One very good qualification was she knew how to handle the young men around the place—they didn't get away with anything." During a trip in the 1960s to Japan Richter exchanged letters with Vi about Seismo Lab operations. Taylor's letters to Richter, and his to her, reveal a friendly, collegial, respectful relationship. Vi sent news from the lab, including comings and goings of secretaries who had a knack for turning up pregnant, and sent regards to both Richter and Lillian. At a time when very few women were involved with seismology in any technical capacity, one has to give Richter credit for what was by all accounts a productive, respectful working relationship women in a professional setting. Towards the end of his career he also worked closely with Karen McNally, the first female postdoc in the Caltech Seismo Lab. When McNally moved on to a position at the University of California at Santa Cruz and became director of the department, she named the university's own seismological laboratory after her mentor.

In professional circles Richter appreciated talent and promoted it, whether it came in a male or a female package. He arguably appreciated women as individuals in his personal life as well. In any case Richter effectively led a double life, and not simply in the usual tawdry sense of the phrase: his life as a scientist and his other life, as a writer. By all indications the other woman were part of the other life: they were fellow writers, in some cases poets, in others, creative writing teachers, and in at least one case, a fellow nudist.

But just how many others were part of his other life, and who they were, we will probably never know. And possibly we never should know. Richter wrote that his notions about relationships were not restricted to those involving only two people. Yet at the same time, while he did not take great pains to hide the existence of other loves in his life, he did go to some length to protect their real names and their privacy. One is inclined to follow his lead. One might argue that the identities of Richter's women don't really matter. One might argue further that their existence doesn't matter either—that these tales represent only so much dirty laundry, the likes of which have no direct bearing on the story of Richter as a man of science. Yet they are clearly relevant to the story of Charles Frances Richter as a man.

In some respects Richter's love life was rather ordinarily complicated. He married relatively young (certainly inexperienced for his years), came to wonder if he had ever truly been in love with his wife, strayed at times (in at least one case seriously), and in the end remained in a lifelong partnership that he came to hold dear in spite of its imperfections. (Simon Winchester blithely refers to Richter as a man of "prodigious sexual appetites." In fact there is no evidence that Richter's predilections were anything but normal for a man whose wife appears to have had predilections of her own.) The possible bond with Margaret, however, suggests emotional turbulence well beyond the usual sorts of storms that sweep through the human psyche.

Yet if his was indeed a life lived largely apart from societal convention, it remained a life of decency, of respect, of integrity. Throughout all of the correspondence to and about the women in his life, colorful though it might have been, any hint of rancor or unkindness—and hint of shirking of responsibility—is conspicuously absent. He might at worst have thought about leaving Lillian during their most difficult years, but he stayed with her, "to carry . . . on in our way; not someone else's way." We will never know the full story of what transpired between Richter and the women in his life, yet perhaps we can understand it. If it is fair to say that complicated people as a rule have complicated relationships, Richter was about as complicated as they come.

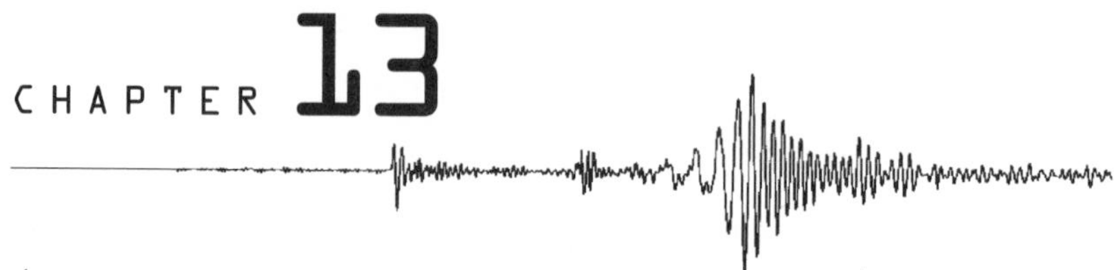

Autumn

My problem is, How can a prevailingly scientific personality like mine find an artistic expression?
—*Charles Richter, journal entry, December 1927*

THE CLUES we have about Richter's affairs of the heart as a younger man are chiefly from personal correspondence, letters written to and received from women who only ever used their first names. His love life remained complicated through his sixties, or became complicated again in his sixties. At this stage of his life he also wrote to the women in his life, the communications not in letters but rather in verse. That he wrote poetry is not generally a surprise among seismologists. Among the handful of personal tidbits about Charles Richter that has been known in the seismological community in the years following his death, is this: he dabbled in poetry. From what one hears one would think Richter penned a stanza or two during an idle moment here and there, or perhaps scribbled a few lines of verse during quiet interludes on hiking trips in his beloved Southern California mountains.

As it turns out, Richter dabbled in poetry the way that Stephen King dabbles in mystery novels—although rather less successfully. His papers at the Caltech archives reveal that Richter wrote—and wrote, and wrote—throughout nearly his entire life. Moreover he wrote not only poetry but also prose: fiction, science fiction, philosophy. His poetry, meanwhile, covered the full spectrum, from stanzas to epics rivaling Milton — in quantity if not quite in quality. The love poems he would write in his sixties in fact represented the culmination of a serious lifelong avocation. The scientist who impressed other scientists as passionate about earthquakes to the exclusion of all else was in fact equally passionate about the facet of his life his colleagues never glimpsed until after his death: his life as a writer.

Judging from both the written work among his papers and by his own account, Richter's career as a writer began following his breakdown in 1921, when psychiatrist Dr. Ross Moore suggested that he try his hand at self-expression in verse. By this time sister Margaret was already a burgeoning poet; it would have been a natural direction for Charles to follow. By his own admission, while he was deeply interested in all art and its role in society, he did not possess the coordination or fine-motor control required for artistic expression apart from that involving the written word. (Even so, one must add, it's a good thing that he knew how to use a typewriter.)

So write poetry he did: poem after poem from spring of 1924, just as he started to find his way back to functionality and to science, until December 1929, although his output appears to have slowed considerably at the beginning of 1926.

Richter's early poetry rambles much as his graduate school journal rambles, tackling similarly weighty themes—nature, God, friendship—with similar emotional fervor. Recall that, by this own account, the demands of a rigorous research program left him "naturally in a cooler and nervously exhausted state," from which his faculty for artistic expression was "blunted, if not entirely lost." It does appears that as his energy for verse diminished, he turned to the more free-form expression of a personal journal. One also notes that he met his wife in 1927, when he was probably a relatively—possibly a completely—virginal twenty-seven years old. Without question he recognized in himself a strong sense of sexual frustration in the years before he met his wife. As we shall soon see, while, by his own account, the demands of his research career modulated his productivity as a writer, the vagaries of a man's love life can surely have the same effect.

In July 1928, within at most a few weeks of the date that he and Lillian were married, he wrote the following:

IN TIME OF PEACE

I did not think to see a day
When I should be as calm as this.
So many stones are cleared away,
I feel that something is amiss.

Have I been granted this brief rest
While destiny prepares a blow?
Must I face battles now unguessed?
I almost hope that it is so

This state of equanimity
Is no less dangerous than strange.
Peace is not natural to me;
My mind is organized for change.

I dread the gradual loss of force
Will take me unawares at length,
If I continue on a course
That does not ask for all my strength.

This is temptation quite as plain
As ever any saint went through.
I must not settle down again;
There is more work for me to do.

Having found his sense of equanimity at last—one can only infer at least in part through conjugal bliss—Richter found himself as unsettled as ever. Having spent virtually his entire life—the span of it that he remembered—in the grip of demons of various shapes and sizes, he hardly knew what to do when they finally went on holiday. It appears Richter understood that what tormented him was inexorably intertwined with the drive that propelled him to accomplishment.

Nonetheless, Richter's writing career does appear to have taken a brief hiatus for a few years following 1927. His papers reveal few poems or other writings dating from this time: what he did write was far less rambling, far more contained. In 1932 he published five short poems in an anthology, California Poets, to which his sister Margaret also contributed; he received a total payment of sixteen dollars.

One poem penned in August 1936 appears to describe the waning years of the honeymoon:

No longer crowded by insistent angels of desire,
Can we not find a calmer space, to watch
 the world and talk?
Because their pounding wings were once so
 swift and so much higher,
Shall the renouncing of such flight
 preclude a pleasant walk?

There may be flashes still, like distant
 lightning when the rain
Is past, and all once sultry air
 is fresh and newly clear.

Even suppose the blinding wings should
crowd us round again,
Would we not find ourselves at last,
sitting and talking here?

Richter had undertaken the Herculean task of developing the magnitude scale during the buoyant early years of marriage. As this pivotal period in his life ended he again poured boundless, often wildly uncontrolled, energies into his writing, in 1937 crafting what appears to be a manuscript on matters of philosophy in addition to his poetry. Echoing his graduate school journal, this manuscript rambles with indiscriminate abandon through serious topics that clearly weighed especially on Richter's mind: the challenge for the introvert to conform in society, ethics, the proper organization of a life, the philosophical implications of the "new physics."

Yet before long his energies were apparently once again diverted elsewhere—and not, it seems, to his research. The previous chapter introduced us to Kim, the young woman with whom Richter quite possibly had his most serious love affair, as well as clues pointing towards an earlier relationship with—or emotional attachment to—Margaret. Echoing the earlier pattern of writing little during his first years of marriage, Richter's journals reveal little poetry written in 1939–40. He did not return to his formerly prolific ways until 1944. He did however pen a love poem or two in the late 1930s, perhaps inspired by his first stirrings towards Kim—or continued stirrings towards Margaret.

One untitled poem from 1938 suggests something more than first stirrings:

To dance behind the shining screen
That cuts off life from half-pretence;
To see and be completely seen
Without distress, without defense;

To be all physical rejoicing
Without one decent reservation,
Your lips, your breast, your toes all voicing
Pure bold delight in animation;

To drop all rules, and let them lie
As if by mere perchance you wore them—
These are your vices, dear, and I
Regret to say that I adore them.

Indirect inference can be as ill-advised in biography as it is in science, yet one can scarcely read these lines without imagining them attached to the dizzying beginnings of a torrid love affair. Richter's writing output might have been reduced considerably when he was swept away by passions of the flesh, yet it was in short love poems such as this that he found the greatest clarity and grace of expression.

Richter's affair with Kim apparently began around 1940; their continuing correspondence, as well as his close friendship with Mavis, carried him through 1945. If his writing output did in fact wax as his love life waned, it appears that both of these relationships had largely run their course by the mid-1940s. By this time he had not only resumed writing poetry full-force, he had also turned much of his writing energy to a novel, "House on a Bridge." This story was, essentially, one long soap opera chronicling the exploits of a small group of friends and lovers (the latter in particular): John, who was first sleeping with Winifred, who was married to James at the time; and Matilda, a young free spirit who became John's second love interest. Reading this story, one can try to pull back the veil on the autobiography: John cast in the role of Richter, Winifred perhaps playing Lillian; and Matilda as Kim. One suspects Richter found catharsis in writing—preserving in fictionalized form the bits he most wished to remember, coming to grips with his feelings of love and loss. Or rather, one hopes he found catharsis. The project did not otherwise meet with success, barely ever coming together as anything that might be considered a coherent draft.

If such was the case, by 1951 the process of catharsis had apparently run its course: by this time Richter's attentions had wandered further, to science fiction. A new novel, "Outlaws on Zem," developed into a full-fledged manuscript that he typed and sent off to John Campbell, an editor with Steel and Smith publishers. In a personally written reply, Campbell wrote that the manuscript was "incomplete," the "plot too complex for careful development. Better material," he added, "results from more detailed development of a relatively simple theme than the loose development of many ideas that you have here."

If there is a single phrase that sums up most of Richter's writings throughout his lifetime, "the loose development of many ideas" would probably be that phrase. His writing, from the diary and poetry of his twenties to his later novels and epic poems, meandered at tortured length over hill and dale, tackling everything and nothing, rarely arriving at answers or even formulating questions with clarity.

This was the man who had developed the Richter scale, an achievement that would not have been possible by an individual less capable of intense, focused, computationally extensive observational science. And yet as much discipline as he brought to this undertaking, no greater was the lack of focus and discipline that he brought to his artistic expression.

Ironically, Richter's artistic expression was at its best when he managed to stifle his tendency towards volcanic outpourings of expression and work on much shorter, simpler works. At these times he apparently felt a keen sense of frustration that his faculty for artistic expression was blunted. Writing was for Richter a cherished outlet, a critical counterbalance to a professional life ruled by discipline, rigor, control. He was happiest when his writing energies had wide room to roam, never realizing, it seems, that great art demands its own balance of freedom and restraint. Even at its best Richter's poetry never rises to the level of great art, yet he achieved a modest level of success with his shorter verse—success that he would not come close to achieving in his more epic endeavors.

During his mid- to late fifties Richter's writing output appeared to wane once again. During this period it seems the demands of his research career rather than his love interests may have been responsible for a diminution of creative energies. These were the years that he devoted to what many regard as the second (if not the first) great contribution of his career: the writing of *Elementary Seismology.* His stepson's death in 1957, and possibly Gutenberg's three years later, may well have further blunted his artistic expression.

Richter returned to writing shorter verse later in life, notably in the late 1960s and early 1970s. During these years his poems again reveal a focus and clarity so desperately lacking in his longer works. And here again, the poignancy of sentiment emerges from the haze of awkward expression.

AFTER READING GOETHE (1969)

Too much
Disturbance now;
The sound of poetry
Is lost in clatter, rumble, roar
And bang!
Wait now;
The evening comes,
When birds in trees are hushed
And over all the mountain tops
Is rest.

NOT SOFTNESS (1970)

At last, I must go now; I can tell where, not why.
The rugged pines and cedars, poking at the sky
A mile aloft, will have the answer. Meanwhile I
Roll up my blanket, store provisions in my pack,
Strap it up, give a jerk to seat it on my back,
And start at sunrise on the zigzag uphill track.
I have clung to the city longer than I should;
Now I must go to wander in the mountain wood.
Kisses and handclasps, apples and roses, do no good.
Wild pines are what I need; their bark is thick and rough,
Even the hillside flowers have the hidden stuff
Of hardness at the base. Petals are not enough.

And, in 1970, Richter penned a soul-baring poem that sounds as if it was written in the twilight years of his life, although in fact, while Lillian's health was failing at this time, Richter himself still had another fifteen years ahead of him:

IN CONCLUSION (1970)

No, I am not ungrateful.
Some living was quite good, and some was not.
Why quarrel with the general human lot?
Not too much has been hateful.

Fear there has been, dark fear
Amid the whirl-wind winds of fear and hate;
Small wonder that I never grew up straight.
Enough; I have survived, I'm here.

Some envy me, but those
Can never know how meager is my part
Of what they take for granted in the heart—
Far less than they suppose

Quietly I descend
These last long stairs, not hesitating much,
Nor fearing that expected gentle touch
That is to bring the end.

One still doesn't worry that Richter missed his true calling by not forsaking scientific research for a life devoted to the arts. At its best, the sometimes touching sentiment of his poetry struggles valiantly to emerge through its awkwardness. He might have had the soul of a poet; what he lacked was the poet's capacity for verbal expression. And yet, had he spent his idle hours cultivating the craft of writing short verses, one can imagine without too much difficulty the happy marriage of science and artistic expression that he yearned for so deeply throughout his life.

In his shorter poems Richter achieved at least a measure of the expression that he so badly craved: expression of a singularly soul-baring stripe. One must take care not to overinterpret Richter's writings: his imagination clearly had free reign, and by his own admission he gravitated towards fantasy and romanticization. Yet it is hard to read his poems and not see expression straight from the heart, reflection on his own demons ("The sound of poetry, / Is lost in clatter, rumble, roar"), internal fortitude ("Petals are not enough"), and love of nature and solitude ("The rugged pines and cedars, poking at the sky, / A mile aloft, will have the answer").

Richter seems to have known from whence he spoke when, as a young man of twenty-six struggling to get his bearings in the world, he wrote, "I think that I have now reached a point at which I can be reasonably certain that my chief artistic difficulty is definitively one of expression and that the problem is not simply that of having nothing to express." That difficulty plagued him throughout his life; ironically and poignantly, he found his voice best at precisely those times when he feared he had lost it.

Thus one is naturally led to wonder about his renaissance later in life: could he have fallen in love again—a man in his mid-sixties who, while never a dashing physical specimen, had achieved professional success and fame? A man who, judging from some of his fan mail, was not altogether unattractive to women? Some of the poems he wrote during these years appear to provide an incontrovertible answer:

For Julia

Women, so men will say, are all alike;
They seem as various as summer bloom
Where the deceiving rays of sunlight strike,
But drop their difference in a darkened room.
I do not know how much of this is true,
Not being deeply versed in woman's way.
Only a few have loved me, but those few
Are not more like by night than like by day.
You I have only known under the sun,
But from your frankness there is much to guess;

Surely within your darkened chamber one
May meet with an exceptional caress.
Yet, if you would, I should be thunderstruck,
For I have never had that much good luck.

An interesting piece of writing, to say the least. The female reader can only cringe at the thought that the first four lines represent an accurate view of what men will say. Setting this aside, one notes the implicit confession—"only a few" is clearly not equivalent to "one." But this we knew, so, moving on, the female reader is heartened to hear him reject the conventional wisdom—if it is indeed such—that women are all alike in the dark. Quickly our relief turns to curiosity: who is this Julia (is Julia her real name?) who has so captured Richter's attentions? Another question: did he write this poem for his own eyes only, in which case the closing lines take on a poignant sense of longing, or for Julia's eyes, in which case they take on an entirely different sense.

Another poem, dated December 8, 1964, reads as an even more explicit proposal:

Why do you hesitate to let me ask
What soon or later you must surely hear?
Is simply saying no so great a task—
Or are you holding back another fear?
Come, let us both agree to shed a tear
Over what might have been; but let us not
Surrender all our joy of now and here,
But rather take what good the fates allot.
All that I need now is an easy spot
To bear the gentle shock of your denial;
You know as well as I precisely what
Moves me unhopefully to make the trial,
But if your answer might perhaps be "yes,"
I should be happier than you can guess.

The questions only grow when one turns the page from "For Julia," dated November 24, 1966, to a second poem with the same date:

This is for you now. I have written much
For Julia and Nerissa and the rest.
Where there is love, wherever there is touch
I write; my paper is a palimpsest
With feelings overwritten and with lines

Crossed out—for love is new and yet the same;
The ancient patterns and the old designs
Group with fresh life about another name.
We are no strangers; we were friends of old,
And this that seems so priceless and new minted
Displays the luster of familiar gold
When seen without enamel and untinted.
Forgive me; if I had not been so slow
We might have owned this treasure long ago.

It thus appears that his desire for Julia went unrequited and perhaps ultimately unexpressed. It would be a bold move, to give a poem such as "For Julia" to a woman by way of invitation. But who, then, was the above poem written for? Who was his "friend of old"? Had he rediscovered romantic love for Lillian after nearly forty years of marriage? Or had an old friend reentered his life in a different role than she had played in his life in earlier years? Could the poem have been written for Margaret? Richter's earlier love letters might raise any number of questions, but they are an open book compared to his soul-baring yet highly enigmatic expression in verse.

Another question springs to mind: who was Nerissa? Richter wrote a great many poems to her during the mid-1960s, poems that leave little room for doubt: "Nerissa! Open wide to me; / In loving you I find reality." That she was a real person is also beyond question. Through the typewritten pages of a binder of poetry one finds small notes, comments on the poems written in pencil, and signed by the same name. That it was not her true name also appears beyond question from following, dated November 19, 1966:

I need a name for you. The one you sign
Is good, sounds well, I could not love you more
Without it; none the less it is not mine.
We need a private name unheard before.
What then? A poet might be Lesbia—no,
Not Sappho, not that bitter love again!
My very dear, I would not have you so;
I want a woman with a taste for men.
Laura, or Juliet, or Heloise—
Which of the lovely visions should yours be?
You might be Beatrice, darling, if you please,
In gratitude for blessings given me.

Nerissa, dear! Nerissa is the name;
This is our legacy to future fame.

As the last chapter discussed, the first half of the poem tends to stop a reader dead in his or her tracks: "*not that bitter love again*"? It is an interesting question—to which bitter love Richter refers in the first half of the poem. One is hard-pressed to come up with an alternative to the obvious explanation. Turning one's attention now to the more immediately relevant latter half of the poem, Shakespeare aficionados will recognize Nerissa as Portia's lady-in-waiting in *Merchant of Venice*, but it remains unclear what clue, if any, this gives us. Or perhaps the name had other significance. At least during the earliest days of the Fraternity Elysia, new members stood before Hobart Glassey to receive a "camp name" drawn from Greek mythology. Nerissa is indeed a Greek name, meaning "sea nymph." (*Julia*, however, is of Latin extraction.) In any case, Richter's Nerissa appears to have been a poet in her own right, perhaps a writing teacher as well.

So, then, we have Julia and Nerissa, and "the rest," as well as Richter's "friend of old." It is not such a stretch to imagine that one or more could have been from the Glassey group: without question he kept in touch with several of them as late as the 1960s. Without question at least one of his earlier loves had been part of this group. Intriguingly, one playfully flirtatious letter was sent by a woman named Ruth on June 14, 1965: teasing that, having closed his last letter with "all my love," Richter must have no love left for any other woman. A June exchange of flirtatious letters might have conceivably have blossomed into a love affair half a year later. But yet again we arrive at possibilities rather than answers. If there is any common thread among the other women in Charles Richter's life, however, it is that were all writers.

A poem dated December 15, 1966, inclines a person to lean towards a late love affair rather than a rekindling of love towards Lillian:

MOUNTAIN DAY

I am awake
With the first bird song, at the earliest dawn,
Watching the peaks take sunlight.
(My love, I think of you all day.)

I make no fire,
But quickly leave my place and take the trail
Following up the river.
(Your feet are delicate and strong.)

The waterfalls
Send rapid streams to cut across the way
And set me little problems.
(With joy I kiss your lips.)

The heat of noon
Catches me on the zigzags up a grade,
Dodging from shade to shadow.
(I worship both your rounded breasts.)

The lake above
Heads in a meadow under scattering trees,
With purple streaks of gentians;
There is my cap.
(My love, I rest in you forever.)

This does not at first blush seem to have been written to a wife of nearly forty years—a wife to whom Richter, by his own admission, felt less than physically romantic love and who may have felt something less than deeply passionate love for him.

Another poem from around the same time, dated November 1, 1964, points even more strongly towards a newcomer:

AUTUMN

No water, earth or air; no fire, not even embers,
Give any habitation to the weary soul,
Fitfully breathing on a dying coal,
Missing the warming glow it wistfully remembers.
Year after year Octobers chill me, and Novembers
Bring actual frost; the later, steaming wassail bowl
Will offer little cheer to me whose only goal
Lately has been to pass another few Decembers.
Had you not come, I do not know if I would ever
Have had the confidence to wait for spring.
You are my spring, yourself; yours is the bluebird's wing
Your gentle fingers, calm and infinitely clever,
Have woven a tapestry where everything
Has reason and a place; your hands have drawn a ring
Around the two of us, that nothing more can sever.

As 1966 drew to a close Richter penned another poem, dated December 30/31, 1966:

Alone, on New Year's Eve! I have no taste
For celebration, meet the coming year
With small regret, but not with any haste;
With reasonable hope, and not much fear.
Companionship would be more welcome, not
Convivial; the touch of hands and not the cup.
A gentle hearth fire in a quiet spot,
No roaring logs, no embers flaring up.
After a lengthy siege of life one knows
Years may be good or bad, and sometimes both.
My middle years were best, but some of those
Began with prayer and ended on an oath.
In silence I shall watch December end,
Content to be assured I have a friend.

Here one returns to a sense of vexing ambiguity. Charles and Lillian spent many of their Christmases apart; it appears that 1966 was one of them. Yet is the poem written to an absent wife whom Richter realizes that he misses very much? Or perhaps to the "old friend" who had reentered his life in recent years? The poem expresses an older man's yearning for quiet companionship: "A gentle heart fire in a quiet spot, / No roaring logs, no embers flaring up."

And then, a poem dated October 10, 1966, but rewritten on January 1, 1967:

After so many years have roared in foam
Over my dike, and cut a channel through,
I turn to you as a forsaken home,
And learn you were more gracious than I knew;
As if I were to turn the rusty locks
Where I had hoarded keepsakes of no worth,
And with the faded violets in a box
Had found a diamond in a lump of earth.
I knew you loved me, knew you wise and kind;
I kneeled and worshipped humbly, gratefully;
But I was young—it never crossed my mind
That you had turned your world around for me.
Now, after all the tumult of the years,
Should I rejoice? Or would you care for tears?

Reading this remarkable—and touching—series of poems, and attempting to decipher their meaning, the scientist can only throw up her hands: insufficient data to draw any conclusions. Here again the historical novelist could draw happily on artistic license to flesh out the details: we would know not only who the old friend is, but that she had auburn curls, a shy smile, and enchanting hazel eyes. The biographer, meanwhile, is left somewhere in the middle, forced to admit that, while we may never have true answers to the questions that these poems raise, we do have enough hints to point to some probable—if not quite the certain—interpretations.

By middle age, Richter's marriage had matured. At the age of forty-nine he had expressed a wish that he could be truly in love with Lillian, but did have deep affectionate feelings towards his wife. He realized that, by staying with her through tough times, he had come to be "genuinely attached" to her. By all accounts she was genuinely attached to, if perhaps not head-over-heels in love with, him as well. Richter had other affairs of the heart, at least one of them serious, in the early 1940s and again later in his life. As for the relationship with Margaret, she is at the same time the one love interest whom Richter identified most clearly, and yet still the single biggest enigma of the larger mystery that was Charles Richter's life.

Through their later years Charles and Lillian continued to spend most of their holidays apart. Her colitis (and perhaps his abiding need for solitude) prevented her from joining his hiking trips in the mountains. His strong homebody tendencies prevented him from joining her on her exotic jaunts to far corners of the world. Did absence make the heart grow fonder?

Letters between the two during his solo summer hiking trips reveal an apparently strong and genuine sense of affection. In summer of 1964 she cursed the medical condition that kept her from enjoying the mountain scenery with him; other letters to him that summer ended with tender closings such as "Be careful, love." He saved the birthday card she gave him in 1965: on the front it read, "I may not be an angel!" And on the inside, "but even a devil may care! happy birthday!" Honestly compels the biographer to add that she signed the card simply, "To Charles, From Lillian." In 1965, however, husband and wife exchanged almost daily letters during her extended trip to France and Africa. In one letter he told her that he would be happier if she did not travel on to Timbuktu, adding, "After all, I do value you." Two weeks later, after she made the trip to Africa and returned to France, he wrote, "never have I rejoiced more than when I saw the French stamp on your letter."

Returning to Richter's own words,

> I knew you loved me, knew you wise and kind;
> I kneeled and worshipped humbly, gratefully;
> But I was young—it never crossed my mind
> That you had turned your world around for me.

It is difficult to read these lines any other way than as they appear—a belated expression of appreciation, contrition, and, if not love, certainly a great depth of warmth, affection, and abiding friendship. But whether those feelings were for his wife or for an old flame that had been rekindled, one can perhaps never know for certain. On balance, his poetry suggests it was not an either-or question: available evidence points to an abiding attachment to Lillian notwithstanding other serious attachments in his life.

Without question, both love and heartbreak shine through a series of poems that he wrote in the months following Lillian's death. Having been in declining health for several years, Lillian's lymphatic cancer was not, according to Richter, fully diagnosed until the fall of 1972. By his account she had felt progressively more unwell through the late 1960s, and had an "internal tumor" removed in July 1970: she had likely developed the intestinal cancer to which colitis sufferers are prone. According to Richter the doctors determined the tumor to be cancerous but did not relay this information to Lillian. She rebounded from the initial surgery and went back to her old life, as did Richter. He left on August 18, 1970 for a hiking trip in Yosemite, returning on September 3. But Lillian's health continued to decline. A lump appeared on her neck in late 1972, leading to more surgery and the diagnosis of lymphoma. She spent ten days in the hospital, returning home on October 25. Only at this time did the grim prognosis become clear. Although she was well enough to return home after surgery, her health soon took a sharp turn for the worse, necessitating a final return to the hospital on November 5. Lillian Brand Richter died at 5:30 p.m. on November 6, 1972, one day before Richard Nixon's landslide victory over George McGovern.

In a notebook of poems from the late 1960s and early 1970s, Richter inserted a single page that read simply:

> Poems afterwards
> Dec. 1972–

The handful of poems that followed are as heart-wrenching as they are short:

De Profundis (December 6, 1972)

The Dark
Fell suddenly;
The twilight was too short.
These stars are not much help; I miss
The sun.

I lie
Trying to think,
Beset by stormy dreams
Not yet accepting what I know
I must.

New day
Will dawn indeed
But with another light. I know
What's lost, is lost; where is the hope,
The gain?

In Peace (December 9, 1972)

Never,
Never at all
To hear that voice again,
To see the face, never to touch
The hands.

Believe,
Whoever can,
Believe that there is more,
And nothing truly ends that once
Has lived.

The worst
Remains, the loss
Not of the joys and hopes
Once had, but deeds not done and words
Not said.

Beyond doubt these verses express the deepest throes of grief. Richter also wrote wrenching letters to Lillian's friends to inform them of her death. Only one of these went to an acquaintance from Caltech circles: ironically,

Hertha Gutenberg. Possibly aware that some in the Seismo Lab had considered theirs to be an odd—and distant—union, he wrote, "We did not feel that we were too dependent on each other—Lillian was a very independent person—yet I feel the loss more than I can say . . . the broken ties, the silent rooms." A palpable sense of numbness and heartache shines through these words, yet one has to wonder how they were received by their recipient. If Lillian had in life been an affront to a woman whose own ambitions and independence had been sublimated according to societal rules of the day, one suspects that these words written about her in death can only have stoked the embers.

And yet, just a few weeks later Richter would pen a pair of poems—their titles linked, typed on the same page—that rekindle the questions all over again:

QUESTIONING (to one)

My love
Cannot decide—
She will and she will not;
And thus she leaves the certainty
To me.
Passion
Blows like a horn
Down sounding from a tower
Arousing armies in the bones
And blood.
One moth,
One butterfly,
Alights in day or dark,
Careless of any thorns, coming
To feast

FAREWELL (to another)

My dear,
You would not care.
Often enough you said
That all of this would signify
Nothing.
So brief,
What yet remains;
To have a little light

To gleam, a little dark and then
Regret.
What lasts
May be a shade,
That you and I must leave,
Bringing our worn-out loves to rest
In peace.

The "Farewell" can only have been written to Lillian: not so much a farewell as a plea for forgiveness following the first poem of the pair. But to whom was Richter's question posed? One suspects it must have been one of his love interests from the mid-1960s, the women he referred to as Nerissa or Julia, or perhaps the "old friend." Intriguingly, in Richter's small daily diaries from the years before as well as after Lillian's death, one finds a number of barely legible entries such as "Lunch, J" (December 2, 1972); "J" (January 10, 1973); and, much later, "Lunch, J, 12:30" (October 24, 1978); "J returns" (May 8, 1979). It appears there was a "J" in his life in the years following Lillian's death. His "Questioning (to another)" was dated December 26, 1972; his date book from that year reads "Lunch, J," on December 28. But who she was, and what, if any, connection she had to the earlier relationships in Richter's life, we can only wonder. (Even the obvious inference that "J" stood for "Julia" is contradicted by the chronology of events as well as what Richter writes in his verse.) It does at any rate appear virtually certain that there were other women in his life in the years leading up to, and quite possibly including, the year of Lillian's death.

Without question, as we have seen, Richter's letters and poems revealed a healthy degree of open-mindedness on matters of romance. His verse reveals unabashed, unapologetic expressions of love to more than one woman, very nearly at the same time. One poem from 1966, a time at which there seemed to be at least two women in his life, perhaps expressed a wish that societal conventions could be something different than what they are:

Household

This is my hope:
To live with doors and windows open wide,
And nothing locked or secret;
A friendly dog
Announcing visitors with joyful barks,
Wagging his tail in welcome;

A graceful cat
Willing to be admired, and perhaps
Consent to purr a little—
And best of all,
To know that other doors will let me in
As lightly as a heartbeat.

Yet conventions were what they were: Richter understood this only too well. Thus are we perhaps left exactly where Richter would have wanted us to be: with an awareness and appreciation for the women he loved throughout his life, an idea of his unorthodox feelings about romantic relationships, and no details that might embarrass the women in his life or their descendents.

About one woman in his life we know a good deal more: for this story we have a beginning, a middle, and an end. Richter not only respected his wife as an individual but also loved her deeply—as evidenced by a poem he wrote in February 1973:

LONELINESS

Crocus;
You set them out
In early spring, last year.
They are in bloom again, and you
Are gone.

The cat
Rests as before
Contented in the chair,
On cushions you laid down for him,
And purrs.

Morning
Brings in the sun
To run his daily race
Over my head, while I remain—
How long?

Richter's feelings shine through these words with clarity, the heartache still raw, its jagged edges unblunted by the passage of three months' time.

"*We did not feel that we were too dependent on each other—Lillian was a very independent person—yet I feel the loss more than I can say . . .*"

The outside world saw a sometimes strained and always unconventional forty-four-year union: it was ironically both more unconventional but also stronger than they knew.

"They reminded me of each other." Brilliant and iconoclastic in equal measure, Charles Richter could not have met his match in just any woman. Lillian Brand was not just any woman. During her final hospitalization, as she lay on her deathbed, a small group of relatives visiting, Lillian suddenly enjoined her husband, "Sing, 'Hallelujah I'm a Bum,'" an old song—a parody of the old Salvation Army tune "Revive Us Again"—thought to have been written by Harry McClintok, who wrote songs as part of the Industrial Workers of the World labor union ("Wobbly") movement. Richter responded simply, "Okay," and indeed proceeded to sing the song as the rest of the family looked on, amazed and touched. They would later guess that the song must have had special significance for Lillian and Charles: one can easily imagine that the philosophy of the defiantly radical Wobblies, as well as their penchant for song and verse, would have had immense appeal for them both. One can also easily imagine they were, if nothing else, interested bystanders of the pivotal milk strike of 1933, which had taken place less than five miles from the Richters' family home in Los Angeles. As Lillian lay dying many years later, Charles made his one last offering to her in verse—then as always, concerned only with what was right, and what was asked of him, caring not one whit what anyone else might think.

Having read Richter's most intimate writings, including those written to other women, we are, ironically, left with a newfound appreciation for the many roles that Lillian Richter played in her husband's remarkable life. For all of the tribulations of their forty-four-year marriage—for all of his wanderings, and perhaps for all of hers—Lillian was, in the end, the woman in whom he found his match; the woman who shared his life; the woman who broke his heart.

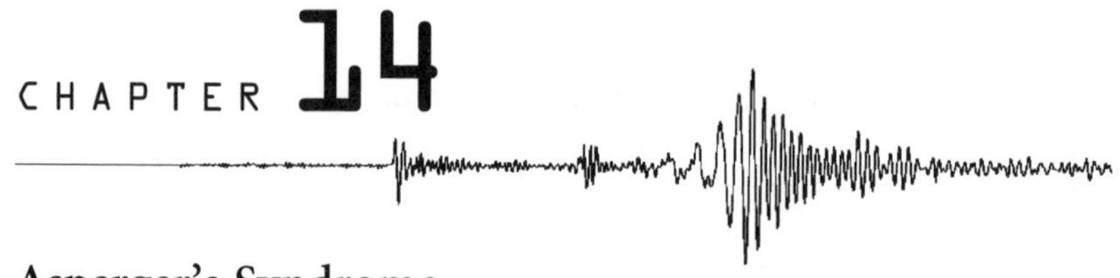

Asperger's Syndrome

The surf is crashing in my brains,
The flooding waves are part of me;
The tides go marching through my veins
No less than on the widest sea.
—*Charles Richter, "Ocean," 1944*

EARLIER CHAPTERS begin to paint a portrait of Charles Francis Richter, the man; to give a sense of the flooding waves that were such an integral part of his extraordinarily talented yet extraordinarily complex personality. Even as he toiled away at some of the key problems in observational seismology—even as he made major contributions, including but not limited to the famous Richter scale—his mind continue to spin tirelessly on the other problems and passions that consumed him. Richter's corporal and intellectual passions were in a league of their own. Many people feel a calling to express themselves through prose; few produce the voluminous outpourings in genres ranging from poetry to science fiction to philosophy. Many reflect on the nature of art and science; few ever devote years of thought grappling with these issues in their minds and in the pages of their journals. Many people feel a passionate calling towards their chosen profession, few have gone so far as to install a seismometer in their living room.

As an aside, one must at some point add a footnote in Richter's defense: In the days before the Internet provided us all, including scientists, with instant communications and access to data, the living room seismometer gave Richter an otherwise unavailable portal to the Southern California underworld. Equipped with a single seismogram and his encyclopedic familiarity with seismograms, Richter could respond to after-hours queries from the media. Among his colleagues Richter was renowned for his ability

Fig. 14.1. Charles Richter with seismometer installed in the living room of his home. (Photo courtesy of California Institute of Technology Archives, reproduced with permission.)

to glance at one or a handful of seismograms and give an estimate of not only a magnitude but also a location. While many seismologists develop some degree of familiarity with seismograms over their careers, few if any have ever matched his degree of intimacy with the pulse of the planet. One further notes that Richter was in good company: Hugo Benioff had previously installed a seismometer in his house after his retirement in 1965. The difference between Benioff and Richter, however, was this: Benioff installed the instrument in his den, while Richter put his in the living room, right next to the grandfather clock.

But I digress. Richter's personality, at least among those seismologists who did not know him in person, has been only vaguely known. More than a few younger seismologists have wondered if he wasn't a bit of a kook—even by scientist standards—and more than a bit of a publicity hound. Such musings have been based previously on the snippets of information that had become widely known: that he dabbled in poetry, that he was an avid nudist at a time when academic decorum did not tend towards the Bohemian. Richter was also known to be so keen to talk to reporters that he kept the Seismo Lab phone in his lap after an earthquake in the area. On a more substantive note was the conventional wisdom that Beno Gutenberg had played an important role in the development of the scale that went on to bear Richter's name alone.

Reading more of Richter's story, especially what he wrote of it himself, one finds oneself led inexorably to a question that nearly answers itself as soon as it is asked: were Richter's follies and foibles the result of a profound organic problem, namely, a neurobiological disorder? The reader might well have reached the conclusion already, so loudly do Richter's personality quirks scream hints in one particular direction: Asperger's syndrome. Named for Dr. Hans Asperger, a Viennese pediatrician, the syndrome was first described in a scientific paper published in 1944, although it is thought to be related to a more profound disorder that has been recognized far longer—autism.

The original paper, "Autistic Psychopathy in Childhood," was written in German and was translated into English only in 1991 by Dr. Uta Frith. Asperger's syndrome is often referred to as "part of the autism spectrum"—one of several so-called pervasive development disorders that share similar qualities. Autism is the extreme end-member of this spectrum: a devastating neurobiological impairment that often leaves individuals unable to function in society, frequently although not always with a severely impaired IQ. In the words of Patricia Romanowski Bashe and Barbara L. Kirby, authors of *The OASIS guide to Asperger Syndrome*, autistic children exhibit "unusual responses to sensory stimuli, seeming lack of interest in other people, abnormally intense insistence on routine and sameness, and unusual attraction to specific objects or parts of objects." (OASIS stands for Online Asperger Syndrome Information and Support, an award-winning Web site from whence the book sprang.) An idiot-savant component of autism, made famous in the 1980s movie *Rain Man*, is only infrequently present, although individuals with Asperger's syndrome sometimes do have remarkable memories for certain kinds of information.

Intensive behavioral therapy, the earlier the better, is usually required for autistic individuals to lead remotely functional lives. Many never do.

Individuals with Asperger's syndrome (AS) are sometimes described as being "a little bit autistic." Children with AS are usually not nearly as obviously dysfunctional as are autistic children, yet they do without question exhibit a marked "differentness" from an early age. As children they can create an enormous sense of frustration among parents, caregivers, and teachers. An AS child is likely to face serious difficulties in school, be perceived as difficult, and have few close childhood friends. Luke Jackson summed it up nicely in the title of a book that, astonishingly, he wrote at the age of thirteen: *Freaks, Geeks, and Asperger Syndrome.* A book written primarily for other AS adolescents, Jackson asks his readers, "Hands up those of you reading this who have been called a freak or a geek or a boffin or a nerd?"

How Asperger's syndrome arises is essentially unknown. It tends to run in families, pointing towards a malfunction deep within the genetic blueprint. Yet a number of studies have reached the vexing conclusion that the incidence of AS has risen markedly in recent decades, suggesting that environmental factors play a role.

A study published in 1998, known as the Lancet study, reached the disturbing and headline-grabbing conclusion that a rise in autism in recent decades could perhaps be attributed to the measles-mumps-rubella vaccine now given to virtually every child in industrialized nations. If the vaccine could lead to autism, then very likely it could account for Asperger's syndrome as well. The findings gained notoriety in the popular press, causing some parents to reconsider what had been routine childhood vaccinations. The medical community as a whole weighed in against the Lancet study, pointing to faulty reasoning, a limited study size, and possible conflicts of interest on the part of researchers. In 2004 ten of the paper's thirteen authors issued a retraction of sorts, stating that their study did not establish a casual link between the vaccine and autism but only raised the possibility. Scientists have pointed to a number of reasons why an apparent link might arise. First, the age at which the vaccine is usually given (fifteen to twenty-four months) happens to coincide with the age at which autism symptoms commonly send worried parents to their doctors. It is moreover possible that mercury, used at one time in the making of the vaccine, could have caused mercury poisoning, the symptoms of which resemble autism. Recognizing the general health hazards of mercury, its use has long since been discontinued in vaccines, although it remains present in the environment to a degree that concerns many health specialists.

An intriguing and frightening twist related to the concern over mercury emerged in 2005. Dr. Raymond Palmer and his colleagues compared the incidence of autism with the amount of mercury in the air and water throughout the 254 counties in Texas. Palmer and his colleagues found a direct correlation: the higher the environmental levels of mercury, the higher the levels of autism. The study, published in the journal Health and Place, found that for every one thousand pounds released into the environment, there was a 61 percent increase in the autism rate. Even the authors cautioned against overinterpretation of the results, which would require considerably more substantiation to be proven. Still, scientists have long suspected that something in the environment may be changing in a way that is causing a rise in autism and Asperger's syndrome. Mercury levels would fit the bill: environmental levels have been rising steadily in recent decades and mercury is a known health hazard in many other respects.

Some have also speculated about a different possible reason for the increased incidence of AS: in short, breeding patterns. Individuals with AS generally gravitate to stereotypical nerd fields: engineering, computers, science, mathematics. When these fields were almost entirely male-dominated, the male of the nerd species was less likely than they would be today to meet, marry, and have children with a female of the same species. Now that women have moved into formerly male-dominated classrooms and workplaces, kindred spirits have more opportunity to meet and produce offspring whose quirky genes are amplified rather than diluted. It is an interesting and plausible-sounding idea, certainly; also one that would be rather difficult to prove.

Until someone finds a way to analyze breeding patterns rigorously, or until the results of the Palmer study are corroborated by larger studies elsewhere, one is left only with faulty genes as a viable explanation for both autism and Asperger's syndrome. And even if a mercury or another environmental contaminant does contribute to the development of these syndromes, there is no question that genetics plays a strong role.

Exactly what goes awry in the brain of an autistic or AS child is also not entirely understood, although scientists have some clues. A 2003 study found that children with autism experienced unusually rapid head growth between the ages of six and fourteen months. Both the head and brain of a severely autistic child are significantly larger than those of normal children from the age of three onward. University of Michigan researcher Catherine Lord suggests the problem may be imperfect pruning: the failure of a child's brain to clear out biological debris as it develops new neural connections. In a 2005 *Newsweek* interview with writer Claudia Kalb, Lord says that in the brains of normal children, "Little twigs fall off to leave the really strong branches"—but perhaps not in those of autistic or AS children. Perhaps it is less figurative than one might think to say that the AS child "cannot see the forest through the trees."

Although the cause and neurobiology of Asperger's are unclear, its symptoms are well established. To anyone who lives with or near an AS individual, the symptoms unite to form a familiar package. One can now consider in detail both the key symptoms, other telltale signs, and other conditions that often exist together with AS. The list follows the OASIS guide, augmented somewhat with additional information from other sources. Although any posthumous diagnosis remains speculative, of course, in the following sections I refrain from belaboring the point that these are inferences based on available data.

More Common in Males

It goes without saying that Charles Richter was male; this point by itself requires no further discussion. Yet the prevalence of Asperger's syndrome among males is interesting to consider. Among individuals referred for a diagnostic assessment of Asperger's the ratio of males to females is close to ten to one—a dramatic imbalance. However, as discussed by noted expert Dr. Tony Attwood, many experts believe that the number should be more like four to one, implying that Asperger's is likely to go unrecognized when it occurs in girls. Attwood advances a number of possible explanations for this. For example, girls are generally more verbal and less aggressive, so an AS girl might be less likely than the AS boy to display inappropriate and disruptive aggressive behaviors. One notes, however, that the casual observation, "Girls are generally more verbal" suggests another, more fundamental, possibility. Given this innate difference, one wonders if AS traits aren't simply less severely expressed in girls because the verbal and social skills of boys do not tend to match those of girls of the same age. A little girl with impaired social skills may thus be "more normal" than a little boy with a similar degree of impairment relative to the norm for little boys. (If this statement offends, talk to any kindergarten teacher.)

Attwood further notes that little girls are often kinder and more tolerant than their male peers. Here Attwood might find himself somewhat at odds with the elementary school teacher who knows that, even by third or fourth grade, little girls can be quite unkind—brutal, in fact—to one another. Novelist Margaret Atwood said it brilliantly in a novel, *Cat's Eye*, addressing this issue: "Little girls are only cute and small only to adults. To one another they are not cute. They are life-sized." From the age of eight or nine, little girls cut their teeth as social animals by forming cliques and engaging in other sorts of games. It would probably be fair to say that, on average, little girls are less of a rough-and-tumble breed than little boys. This, perhaps in combination with girls' stronger social skills, might lead to an easier acceptance of the AS girl by her peers than the AS boy among his peers. As Attwood notes, an AS boy will often prefer the company of girls to other boys. A further distinction between AS girls and boys is that girls' special interests may be focused in less antisocial directions: the AS girl might be passionate about horses or literature rather than electronics or computers.

Charles Richter exhibited many of the stereotypical symptoms of AS as expressed in boys. Born far too early to grow up with computers, Richter turned his attentions to the stars, and then to mathematics and science. One suspects that, had he been born fifty or so years later, he would have been a world class computer geek. Indeed, when a Caltech professor invested several thousand dollars to buy one of the first desktop calculators, Richter pounced on the opportunity to put it to use. (One shudders to imagine what his writing output would have been like with modern word processors at his disposal.) In any case, if AS individuals are male, most males do not have AS. The following sections consider somewhat more diagnostic traits and behaviors.

Difficulty with Social Relationships

There is little doubt that this was true of Richter's life, starting in early childhood. His difficult family circumstances no doubt contributed to an extreme degree of social isolation: nowhere in any of his writings does he allude to friends of his childhood, or even of his later youth. Still, children are naturally social animals who even under difficult circumstances find an easy sense of kinship and camaraderie with their peers; they understand social dynamics as they understand how to breathe. Yet by Richter's own account in 1945, it was his 1909 introduction to formal schooling that brought his first contact with other children and the first real troubles in his life. He had been left largely to his own devices up until that time, a situation that left him quite content—even happy—rather than lonely.

Normally social young children can be dropped into settings where they don't speak the language and will nonetheless thrive after a passing phase of adjustment. Hertha Gutenberg recalled that her nine-year old son was quite unhappy during his first months in California: fearing that the boy would not be able to keep up in classes taught in English at his own grade level, the school had placed him in first grade. A stranger in a strange land, and among much younger children, Arthur Gutenberg frequently came home from school and dissolved into tears. Still he learned English quickly, and was back with his rightful classmates two months later. His younger sister reportedly never spoke a word in English for six weeks, even as she mingled with classmates and playmates in her neighborhood. After six weeks, according to her mother, "All of a sudden she spoke English." Within two years both children rejoiced upon their return to California after a visit to Germany. This example typifies the stories one comes across not infrequently in day-to-day life: the child from the People's Republic of China dropped into a fourth-grade California classroom not speaking a world of English, the American child plunked into an Italian school speaking only a handful of words of Italian, the young refugee child from a war-torn nation who suddenly find himself in a whole new world. Children are social creatures with an enormous capacity for learning; they bloom where they are planted, often with astonishing speed and ease.

In contrast, by Richter's explicit account, it was only among follow nudists in the Glassey group that he found his first real sense of friendship—at the age of thirty-five. It thus seems that he found this first friendship among a group of like-minded iconoclasts. For individuals with AS, social interactions represent a monumental, never-ending challenge: everyone else except you, it seems, knows the secret handshake that paves the way for normally easy relationships. Clare Sainsbury's book, *Martian in the Playground: Understanding the Schoolchild with Asperger Syndrome*, argues that the AS child is like (and feels like) a Martian in a world of earthlings he cannot relate to. An AS child might unintentionally bully other children, or become the target of bullies who learn that the child will respond inappropriately to provocation. Not surprisingly, older AS individuals, for whom the lack of social interaction might well have become a source of consternation, are likely to feel more at home when they finally find themselves among fellow Martians. With effort and advancing maturity, they can also learn skills that do not come naturally. In 1945 Richter wrote, "Some men may be so fortunate to have a natural talent for living; with most, like myself, it is an acquired art, learned mostly from women."

Like-minded iconoclasts did finally appear in Richter's life. Although they may have seemed few and far between, their numbers began to add up: members of the Glassey group, his nuclear family, his wife's family, his stepson, fellow members of the book club he and Lillian belonged to for many years, the other women in his life. Still, even later in life, Richter generally impressed people as simply *odd*. Recall the impression that Richter made on two strangers with whom he and Lillian shared their 1937 mule tour of the Grand Canyon. "At dinner," Gladys Broderson wrote, "we noticed what a peculiar person 'Charles' was—he never looked at you when he spoke and just sat with a grin on his face that seemed to indicate he was in on some joke that you were not." And yet Broderson appears to have been not only a delightful essayist but also a perceptive observer of people: "Charles was so slow and pokey we didn't know who would hire him—maybe he just caught butterflies for a living, but then we [decided] that he was too slow for that so Charles['s] means of earning a livelihood remained a mystery to us." (One wonders if she ever learned the answer!) "However," she continued, "we were grateful to Charles for having handed us a lot of laughs and maybe 'Sweet-thing' (that was what Lillian called him) has a beautiful Soul."

A few of Richter's closest colleagues came to know and appreciate the soul beneath the eccentric exterior. In a letter read by Clarence Allen at Richter's memorial service, Frank Press described Richter as "the most unusual colleague [he] ever had," going on to describe the "natural grace and humanity that was evident despite external social awkwardness." At the same ceremony, Eric Lindvall relayed his surprise at receiving a letter from lawyers with whom Richter had interacted in his capacity as a consultant. Although Lindvall had been under the impression that lawyers as a species had never been particularly fond of Richter, the letter described the "delight to be his presence . . . to note his warmth and delightful sense of humor. He endeared himself to our hearts and we share your sense of loss."

Richter also clearly endeared himself to women's hearts. "Only a few," he wrote, but a few who clearly cared about him deeply. Considering the extent to which interpersonal relations challenged Richter's impaired social skills—later as well as early in his life—one is almost inclined to feel impressed that there *were* other women in his life. One is certainly inclined to feel impressed with the number and quality of friendships he made, and kept, throughout his life.

Sensitivity to Light and Sounds

Day-to-day life presents a never-ending assault on the sensibilities of those with Asperger's. In addition to difficulties with social interactions, AS individuals are often highly sensitive to loud noises or certain kinds of noises, regardless of volume. For example, the infernal gas-powered leaf blower might be a minor annoyance to most people yet cause the AS individual to want to crawl out of his skin. Certain kinds of lights can pose a problem as well: strobe lights, flickering lights, fluorescent lighting. As Liane Willey writes in her book *Pretending to be Normal: Living with Asperger's Syndrome*, "Together, the sharp sounds and the bright lights were more than enough to overload my senses. My head would feel tight, my stomach would churn, and my pulse would run my heart ragged until I found a safety zone." Bright lights bothered Richter as well. According to nephew Bruce Walport, Richter once refused to proceed with a television interview unless the bright lights were turned down. Informed, "This is television, we have to have lights," Richter replied that either the lights had to go or he would. The interviewer relented, the lights went off, and Richter continued the interview as a disembodied voice in the darkness. Noise bothered Richter as well. In a 1954 journal entry he wrote, "Of course, I cannot write or even read poetry against noise, still less against noisy music—and in our home, our neighbor's radio . . ."

Liane Willey describes the solace she found underwater. Richter found his sanctuary in the mountains, where he spent weeks at a stretch hiking on his own, sometimes in the San Gabriel Mountains just north of Los Angeles, often in the Sierra Nevada. The wilderness offered Richter an escape from everything: not just bothersome light and noises but also the social interactions that posed such a consternation for Richter throughout his life.

Richter himself wrote, in one of his few poems from 1926,

> Dear mountain, do not cast me off so soon;
> I come for help. Down in the distant street
> I live through daily bruises and reproof.
> The mountain things can never know how sweet
> The wide air seems, the quiet and the height,
> The half-caressing cleanness in the light.
> After the town I even love the fog,
> And bless the very stones that hurt my feet.

Noise and bright lights were clearly part and parcel of the onslaught of everyday life that provided his "daily bruises and reproof." By his own account in 1945, the rugged mountains of California provided an emotional anchor even in early adolescence.

Kate Holliday reached precisely this conclusion after spending time with Richter for the purpose of writing an article, "Backpacking: The Ageless Art," for a 1973 issue of *Field and Stream*. "Talking with him," she wrote,

Fig. 14.2. Photo published with *Field and Stream* article "Backpacking, the Ageless Art," 1973. (Photo courtesy of *Field and Stream*, reproduced with permission.)

"one gets the impression that, even at 72, the world is too much with him, in the laboratory, in the sudden, urgent call for his know-how . . . in his long fight for public and private construction which will withstand . . . upheavals and thus save lives." She wrote further, "His wanderings are of the spirit, not the body. They feed his need to reassess his own philosophy of both man and nature, to join them once again in ancient, simple harmony." The mountains provided Richter with a respite from the environmental and societal drone that screamed literally as well as figuratively in his ears and glared brightly in his eyes.

Extreme Preoccupation with a Special Interest

This barely requires elaboration. The question is, however: when does interest in a hobby or profession cross the line and become kind of overwhelming preoccupation that is one of the key hallmarks of Asperger's syndrome? According to the *OASIS Guide*, "Immersion in the special interest may come at the expense of other, more socially appropriate activities."

It is clearly a matter of degree, whether a special interest is part of an otherwise balanced life or becomes all-consuming. Anyone who has spent time with AS children knows that they don't do anything halfway. And "adults with AS are more likely to read about their special interest than to talk about it. A special interest 'exchange' is noticeably one-sided, run-on, and more of a monologue than a conversation." AS children are sometimes described as "little professors."

Richter never wrote at length about his early childhood, although he did speak of an early fascination with astronomy that was passionate enough for him to make lengthy, detailed, and useful amateur observations. And once he turned his adult attentions to something, he did so with the hurricane-force intensity. One of the most unfortunate—or at least one of the most complicating—aspects of Richter's life is that his interests focused on more than one passion. He yearned desperately for artistic expression even as he felt himself drawn inexorably to pursue his more natural talents and interests in the sciences. Compounding the problem further, he fixated not only on the two very different passions but also on the need to reconcile what he saw as inconsistencies between them. A more balanced individual might simply do research during the day and write poems, or play banjo with a bluegrass band, on the weekends. By virtue of his unique brain wiring, Richter's passions could not be balanced comfortably. It is, then, small wonder he nearly tore himself apart as a young man. Small wonder he produced stacks of poems and manuscripts on evenings and weekends even as he authored and coauthored stacks of scientific manuscripts that represented seminal contributions to the field.

Richter's extreme preoccupation with seismology—acquired later in life if not as a young man—is legendary. According to former colleague Jim Whitcomb, Richter remembered the birth years of many scientists by virtue of what was for him a simple trick: he associated birth years of individuals with notable earthquakes, the dates of which he never forgot. Quoting again from the OASIS guide, "'Walking encyclopedia' is often an accurate description of the amount and depth of information children with AS accumulate." On the occasion of Richter's death in 1985, Don Anderson, then director of the Seismo Lab, observed, "He loved earthquakes, and he was a walking encyclopedia of seismic data."

Stiff, Pedantic, One-Sided Conversations

This trait is part and parcel of the tendency towards extreme special interests. Charles Richter might have started out as a reluctant seismologist, but eventually his adopted field became an adopted passion. We do not know what Richter's childhood conversations were like, but we have heard about his conversations later in life, in particular on the subject of earthquakes.

In the 1950s CBS interview, Richter's faltering voice never conveys a sense of real authority, but it does find itself more resolutely on track when he speaks specifically about matters of science. Richter considered himself to be a poor instructor and speaker, and although he was tapped later in life to give many lectures to distinguished groups, he could have had far more speaking engagements had he sought them out. He avoided scientific meetings as well, attending far fewer than most of his colleagues. But if he did not like to lecture to groups, he certainly loved to talk at exhaustive length to individuals, in particular to reporters who called and visited to ask questions. Richter probably would have frustrated the socks off of journalists of the present day: the exigencies of modern news reporting rarely allowing room for more than a catchy sound bite. As the CBS interview makes clear, however, journalists of earlier times had the luxury of more in-depth interviews—asking questions to which long, even pedantic answers were not unwelcome. What his colleagues saw as grandstanding emerges as something else: in this one setting, Richter was in his element.

Kate Holliday summed it up thusly in a 1971 article in Smithsonian: "His talk is simple, almost aimless, until he begins to speak of his own particular field. Then his words take on such unblemished authority that the listener immediately gets the feeling that Richter knows, and that all he is doing is nibbling around the edges of that vast knowledge."

Richter's penchant for pedantic, one-sided conversations also shines through with crystal clarity in his 1979 interview with Ann Scheid. When the questions veered towards the personal, for example regarding his wife, the answers are so short and the subject changes so quickly that it seems as if interviewer sensed a level of discomfort and moved on. Richter is by far the most long-winded, supplying answers that stretch to one or more double-spaced pages, when the questions are of a purely technical nature: "Do you think that the new sources of information are going to be helpful in prediction?" "So maybe each fault has its own pattern [of earthquakes]?" And so forth. And so on.

Apparent Lack of Empathy

"Apparent lack of empathy" is a trait that others would notice much more readily than an AS individual—even an unusually perceptive one—would recognize in himself. This trait is an aspect of Asperger's that is part and parcel of an overall package—one that leads to pervasive difficulties due to the frequent disconnect between an AS individual's intentions and how his behaviors are perceived.

Consider, for example, Richter's dispassionate answers to interviewer Ann Schied's questions about women in the early days of seismology. His terse replies gave the impression of a man who had not given the matter much thought, much less any real concern. And yet the role of women in a changing society was among the weighty issues that he grappled with as a younger man; almost surely these issues continued to weigh upon his mind in later years. This example illustrates the disconnect, in particular the "apparent lack of empathy": a person who in fact cares a great deal might well give the impression of not caring at all by virtue of difficulty with verbal expression. In Richter's case it would be more accurate to say oral expression: while he did not possess the gift of poetic expression that he yearned for, he clearly expressed his innermost thoughts through the written word far more easily than the spoken one.

Many of Richter's actions might well have been interpreted by his colleagues as a lack of empathy. By his own admission he struggled mightily against a tendency to procrastinate. In academic circles, this would equate to papers and reviews turned in late, a failure to uphold committee responsibilities in a timely fashion, and so forth. When an editor or committee chair is left twisting in the wind, his own work on hold in the interim, he might well curse the blasted insensitivity of the individual responsible.

Here again we can turn to the words of one of Richter's longtime colleagues to illustrate the point. Following Richter's death, Robert Sharp observed "He was very compassionate about the people living in those buildings," Sharp said, referring to older buildings that were known to be unsafe in earthquakes, "and he had a strong sense of community with people. He just had trouble expressing those feelings."

Problems with Social Use of Language

Language difficulties are, in a sense, a specific component of the larger problems that AS individuals have with social interactions. Language is, after all, a critical key to relationships of any sort. Most humans communicate with each other naturally; those with AS struggle with spoken language in particular. Individuals with AS fail to grasp nuances and symbolic use of language even when they understand word meanings and sentence structure; they struggle when word meanings depart from the literal or the specific. As discussed in the OASIS guide, parents of AS children find themselves frustrated when they tell the child to "get ready for school" and the child does not immediately grasp the many implicit instructions: get dressed, eat breakfast, collect your belongings, and so on.

One might wonder how an individual with AS could write even halfway respectable verse, since symbolic use of language is a cornerstone of poetic expression. One answer is that Richter's use of symbolic language in verse was not overly successful: his most effective poems were literally soul-baring rather than rich with symbolic imagery. Richter himself provides another answer, in a draft of a letter dated September, 9, 1936. One cannot tell to whom the letter was written, or if it was sent at all, but it appears to be a fumbling expression of interest and affection to an individual who one can only assume is a woman. From the timing, the recipient could have been "Kim": the woman he writes to in 1941 after what was clearly a serious affair in the preceding years. It could also have been Margaret.

It is difficult to interpret a letter without knowing to whom it was written, but one passage stands on its own: "I find that I have run full tilt into a group of facts with which I was already familiar in another connection:—namely, the difference between the intellectual use of words to define, denote and express concepts—and the poetic use of words to suggest, to call up feelings, to behave as signs and symbols of that which cannot be expressed directly. What I am not accustomed to is the occurrence of the second, the symbolic function of words, between individuals. I feel that I have been a little obtuse—I should have realized before that this is the explanation of those occasions when I have felt that much more in the way of mutual comprehension had passed between us than could be accounted for by the meanings of the words that had been said."

Thus does Richter describe with some precision one of the key hallmarks of Asperger's syndrome: the inability to understand figurative use of language, such as symbolic representation, and nuances, *in a social context.* What made him unusual among individuals with AS was that he had an intellectual understanding of not only the role of symbolic language in artistic expression but also its role—one that vexingly eluded him—in interpersonal communications.

Richter exhibited other typical language-related traits of AS. Because individuals with Asperger's take words at face value, many struggle with the concept of humor. Richter fit the bill so well that it became legendary in scientific circles, his inability to take a joke, in particular to laugh at himself. He was not without a sense of humor; wry observations can even be found among the pages of Elementary Seismology. In a footnote on page 28, for example, Richter relates the tale of the astronomer who discovered that the local army post fired a cannon at noon every day, using a clock in a jeweler's window for their own timekeeping. The jeweler in turn explained to the astronomer that he set this clock by the firing of the local cannon at noon.

Yet as we have seen, Richter bristled with fury if his colleagues dared to laugh at the perceived joke of showing up wearing two ties. One especially renowned vignette played out at his retirement party in 1970. Colleague Kent Clark had earlier composed "The Richter Scale," which was sung by colleagues at the party. The song began,

> One two on the Richter scale, a shabby little shiver,
> One two on the Richter scale, a queasy little quiver . . .

and worked its way all the way up the scale:

> Someday pretty soon we fear our many faults will fail us
> Slide and slip and rip and dip and all at once assail us
> Seismic jolts like lightning bolts will flatten us that day
> When the concrete settles down geologists will say, it measured
>
> Eight nine on the Richter scale, it rocked 'em in Samoa
> Eight nine on the Richter scale, it cracked like Krakatoa

Richter's fury reportedly rose right alongside the magnitudes in the song, and rose especially when it was sung a second time. Weeks later he told friend and colleague, Robert Sharp, who tried to soothe his feelings over the incident: "My science," Richter fumed, "is *not* a joke."

The Geologic Division at Caltech has a long tradition of year-end parties where faculty, students, and staff get together to eat, drink, and have fun, largely in the form of skits that poke fun at one another. So-called Zilchbrau parties continue to this day, a chance to cut loose in an organization where the business of the day is scarcely unfriendly but often intense. During one such gathering in the early 1960s a skit featured a seismologist who took off his stain-splattered tie and used the distance between spots to determine an earthquake magnitude. Nobody in the audience that day had a shred of doubt whose ox was being gored; nor did they have any doubt that at Zilchbrau, oxen are only ever gored with respect and affection. Such merriment was not, however, Richter's cup of tea. The *Los Angeles Times* obituary quoted one Caltech professor as saying, "Charlie came to a few

THE RICHTER SCALE

Charley Richter made a scale for calibrating earthquakes
Gives a true and lucid reading every time the earth shakes
Increments are exponential, numbers 0 to nine
When the first shock hit the seismo everything worked fine, it measured

One two on the Richter scale, a shabby little shiver
One two on the Richter scale, a queasy little quiver
Waves brushed the seismograph as if a fly had flicked her
One two on the Richter scale, it hardly woke up Richter

Nineteen hundred thirty three and Long Beach rocked and rumbled
School house walls and crockery and oil derricks tumbled
Hollywood got hit but good, it even shook the stars
Shattered glass and spilled martinis on a hundred bars, it measured

Six three on the Richter scale, it rattled tile and plaster
Six three on the Richter scale, a rattling disaster
Waves bounced the seismograph as if a cue had clicked her
Six three on the Richter scale, it almost rattled Richter

Came the turn of County Kern, the mountains lurched and trembled
Bakersfield, which jerked and reeled, was almost disassembled
Arvin town was battered down in rubble and debris
Spasms racked the women's prison at Tehachapi, it measured

Seven eight on the Richter scale, it fractured rails and melons
Seven eight on the Richter scale, it fractured female felons
Waves smacked the seismograph, a casualty inflicter
Seven eight on the Richter scale, it almost fractured Richter

Came a cataclysmic quake at Anchorage Alaska
Seisms ran from Ketchikan to Omaha Nebraska
Polar bears were saying prayers, the tidal wave was grand
Planted boats in California way up on the sand, it measured

Eight five on the Richter scale, it loosened kelp and corals
Eight five on the Richter scale, it loosened faith and morals
Waves bashed the seismograph as if a mule had kicked her
Eight five on the Richter scale, it failed to loosen Richter

Someday pretty soon we fear our many faults will fail us
Slide and slip and rip and dip and all at once assail us
Seismic jolts like lightning bolts will flatten us that day
When the concrete settles down geologists will say, it measured

Eight nine on the Richter scale, it rocked 'em in Samoa
Eight nine on the Richter scale, it cracked like Krakatoa
Waves crunched the seismograph, just like a boa constrictor
Eight nine on the Richter scale, it really racked up
One two on the Richter scale, three four on the Richter scale
Five six on the Richter scale, seven eight on the Richter scale
Eight nine on the Richter scale (CRASH)
It really racked up Richter

K. Clark

Fig. 14.3. Lyrics to song "Richter Scale," written by colleague Kent Clark, and performed at Richter's retirement party in 1970. (Courtesy of Caltech Seismological Laboratory, reprinted with permission.)

of these parties before he finally came up to me and said, 'Frankly, I can't take this sort of thing.' He never attended another."

A Tendency for Motivations and Actions to Be Misunderstood

The behaviors and actions of individuals with AS are all too easily misunderstood, as earlier sections have illustrated. A person might appear to have no sense of humor when in fact he struggles only to appreciate certain types of humor. A person might appear hopelessly mercurial, even childish, when in fact he is struggling valiantly to cope with overwhelming exigencies of daily life.

By Richter's account following the retirement party incident, "There [had] always been too much comedy introduced at Caltech on what should be serious occasions." Yet two documents reveal how, here again, Richter's simple "inability to take a joke" (not to mention his temper) was not what it seemed. In a letter to Sharp, Richter expressed his dismay that the song had created a "farcical atmosphere." He went on to say, "I had in mind some serious and carefully prepared remarks—elder statesman stuff, for the benefit of the younger staff and students. When, later, I offered some of the mildly humorous introduction which I intended, I was met with such guffaws after every word that I could scarcely make myself heard. So I gave up."

A typewritten two-page document reveals the speech that Richter had apparently intended to give: thoughtful and poignant, including a poem by Tennyson and thoughts on warfare that ring dishearteningly true with to the modern ear. He wrote

> Lately, I have been shocked repeatedly by those who insist that we who lived through those two catastrophes should therefore be sympathetic to the officially sponsored side of the conflicts that are now going on. I feel that I have learned better than that. It is painful to hear the same arguments now that were used to justify the two great wars—and along with them the same half truths, and probably some of the same lies.
>
> Naturally, young people notice some of this. They are restless and angry because they know they are being lied to, even though they cannot get at the truth. Some of these overlook that all the lying is not being done by one side; there is plenty of outrageous lying put forward by so-called revolutionaries.

> In spite of the greatest development of information services the world has ever known, I think that no ordinary citizen has any real access to the facts on which he might base sound judgement on the national and social issues of our times.
>
> I dislike many things that Spiro Agnew has said; but some of his remarks about the news media were justified and deserved. I speak from personal experience, over many years, with the manner in which news is handled.

The speech continued from there to address issues of more parochial concern, including an allusion to long-standing debate about whether the Seismo Lab should occupy itself chiefly with global or local earthquake problems.

As Richter wrote to Sharp, he did not view retirement as a happy occasion: "Getting old and being shoved on the shelf is not funny," he wrote, "in fact, it is just plain hell." An intensely private man among his colleagues throughout his entire career, Richter had planned to mark this auspicious, if not entirely happy, occasion by sharing some of his innermost thoughts with students and younger colleagues in particular. Instead, he found his audience in a mood to eat, drink, and be merry.

The tale that would be told (and it surely was told) by successive generations of seismologists following the party painted the picture of the man Richter's colleagues thought they knew: passionately and exclusively interested in earthquakes, famously unable to laugh at himself. It seems only fair to give Richter his due within these pages, the platform that he did not have in 1970: in the appendix the reader will find the full text of the speech that Richter planned to share on the occasion of his retirement, not too much the worse for wear after thirty years in a box. It is, one imagines, safe to assume that, had the speech been given as planned, it would have surprised colleagues who had hitherto never glimpsed their colleague's deeply philosophical side. The intentions of people like Richter are so easily and often misinterpreted by people for whom AS individuals are as inscrutable as the world is to an individual with AS.

Richter remained inscrutable indeed throughout his long career, up to and beyond the day that it formally ended. Emile Okal, a Seismo Lab student in the late 1970s, perceived Richter as very much addled in his later years: the man rarely spoke, and, as Okal discovered one day, came in to pick up mail that then languished unopened in stacks in his office. The ravages of Alzheimer's had, it seemed to Okal, taken an enormous toll on the man's mind. Indeed, by his late seventies Richter had pulled away from the relentless day-to-day routine that he had struggled with throughout his career. But those letters that he did write, as well as interviews he gave into the early 1980s, reveal a man in impressive possession of his memories as well as his marbles. In February 1981 he wrote a letter that he closed with a few thoughts on journalists, "Over many years, I developed respect and sincere sympathy for experienced journalists; but some of the lesser lights of the profession can be annoying, as you seem to have found. I have sometimes used a quote from the great Samuel Johnson (in Boswell's Life, under the year 1784): 'Johnson having argued for some time with a pertinacious gentleman; his opponent, who had talked in a very puzzling manner, happened to say, 'I don't understand you, Sir;' upon which Johnson observed, 'Sir, I have found you an argument; but I am not obliged to find you an understanding."' Although the letter reveals that the typewriter, like the typist himself, was aging, it also reveals nearly perfect spelling and punctuation.

Karen McNally, the last Seismo Lab colleague to work closely with Richter, says with certainty that he did not have Alzheimer's. His mind and his humor remained biting, quick, and intelligent throughout McNally's years at the lab, which ended in 1982 when she took a faculty position at UC Santa Cruz. Even in 1984, when Richter was confined to a convalescent home, suffering heart problems as well as painful shingles, and frequently withdrawn, he would become animated, talkative, and lucid when she arrived to visit. Clarence Allen also found Richter quite lucid during his visits to the convalescent home. He did, however, note his colleague's tendency toward "ever-more cantankerous behavior."

Motor Clumsiness

Here we get into traits that, while perhaps not key diagnostic symptoms, represent what the OASIS guide calls "other tell-tale signs" of Asperger's. "The majority of those with AS struggle with some degree of fine and gross motor skill deficits." "Many . . . have a great deal of difficulty with handwriting." Here again there is no room for doubt that Charles Richter fit the bill: one glance at his journal pages provides all the evidence one needs regarding his penmanship—or lack thereof. He was clearly physically awkward and not well coordinated: by his own account this led to his poor showing in chemistry at Stanford, and led him to switch to a field of science that did not involve pouring sometimes hazardous chemicals into sometimes fragile beakers.

As for his handwriting, his papers reveal that, when he worked very hard at it, his penmanship achieved a level of proficiency that could best be characterized as marginally respectable. When he didn't work very hard at it, for example in the journals and other writings that he wrote in small notebooks during his hiking trips, the result was scribbles so inscrutable that one marvels, truly, that he was able to read his own handwriting. (Clearly he was able to read his own writing, as evidenced by the fact that he typed many of his reports once he returned home.)

Problem with Organization

Quoting again from the OASIS guide, "Asperger Syndrome compromises executive function, or the brain's ability to plan and carry out the steps to complete the task or behavior at hand." Reading through this and other passages, one cannot help but wonder how Richter would have responded, had he had a chance to read them himself. One suspects it might have led to an epiphany of earthshaking proportions, the sense of understanding of self that had so long eluded him. He did understand himself well enough to paint a nearly textbook portrait of individual symptoms. The lack of organizational skills he described as a strong tendency towards procrastination: "There must be a dozen research papers for which I have started assembling material at one time or another, but which have never been finished because the work was postponed in favor of something else." This he wrote to Dr. Moriarty in 1949, as part of his discussion of his proclivity to procrastinate badly unless he imposed extremely rigid habits on himself, as he did for the routine laboratory work that had to get done.

Elsewhere, his limited organizational skills worked, by his own admission, to his disadvantage: "I have noticed that, like this committee program, the activities I tend to postpone are such as would contribute to my personal prestige and reputation." He went on to talk about a different issue, essentially a fear of failure, but he goes on to talk about similar dilatory behavior in other aspects of his life, such as preparing his income taxes and paying bills. "As a youngster," he wrote, "I used to go a day or two without shaving. Even now I am not very reliable about baths, care of the teeth, clean clothes, and so on." One suspects this did little to ameliorate his difficulties with social interaction. Nor was he fastidious about his surroundings. Kate Holliday described his bedroom as cluttered, "like a teenager's."

One of Richter's colleagues, when presented with the notion that Richter had Asperger's syndrome, questioned whether an individual with profound organizational difficulties could have written a book like *Elementary Seismology.* The organizational effort required to put such a book together would indeed have sorely taxed a scientist who struggled to put together even short papers. It appears that Richter was able to harness the horses, so to speak, for this major undertaking: the crafting of his magnum opus. The effort may have moreover left Richter utterly spent. In his letters from the late 1950s onward he sometimes wrote of "nerve strain" or health or mental health difficulties. His scientific contributions were especially sparse between the mid-1950s and mid-1960s. One finds, moreover, remarkably few poems written during this time—or love letters. (The death of his stepson in 1957 may have further contributed to his emotional struggles in the late 1950s and early 1960s.)

"Lack of Focus"

Perhaps part of the overall difficulties with organization, Karen Williams from the University of Michigan sums up focus problems as follows: "Children with AS are often off-task, distracted by internal stimuli; are very disorganized; have difficulty sustaining focus on classroom activities. Often it is not that the attention is poor but, rather, that the focus is 'odd.' The individual with AS cannot figure out what is relevant . . . attention is focused on irrelevant stimuli." She further observes that AS children "tend to withdraw into complex inner worlds in a manner much more intense than is typical of daydreaming, and have difficulty learning in a group situation."

Individuals with AS, and probably to an even greater extent, those with autism, in some ways suffer from an overabundance rather than a lack of focus. Most people perceive the world through strong filters that block out the otherwise overwhelming onslaught of information and stimuli. Individuals with autism or Asperger's syndrome finds themselves paying attention to any and all details—and completely overwhelmed by sensory overload.

Here one perhaps arrives at last the key to Richter's voluminous, hugely ambitious but hopelessly unfocused writings. In many of his writing efforts Richter combined all of his AS traits with his prodigious intellect and boundless energies. The AS trait of poor focus combined disastrously with his multitude of interests, essentially leaving him with a three-hundred-horsepower engine and a transmission that slipped madly between gears. On an intellectual level he yearned desperately to tackle fundamental problems, not only of science but of philosophy, art, and other subjects. He yearned further to reach the pinnacle of artistic expression through poetry and prose. And yet by virtue of his innate makeup, he set off to explore the grandest forests known to mankind and found himself hopelessly mired in a thicket of trees.

The same sort of problem appeared to have plagued his research as well. He produced research papers either with prodding from a highly organized task master, or focusing on his own short, limited investigations that could be completed and written up in short order. If he embarked on more complicated, involved studies they often languished unfinished. His short, highly focused studies were in a sense equivalent to the short, focused poems that he was sometimes able to craft—poems that succeeded at least in small measure whereas his more ambitious writings failed miserably. This starts to sound very much like what Karen Williams had in mind: "it is not that the attention is poor but, rather, that the focus is 'odd.'" In his writing as in his research, Richter struggled to find and maintain an appropriate degree of focus: to get his brain into the proper gear, and keep it there.

Difficulty with Transitions, Changes in Routine, Surprises

This AS trait may stem directly from the former one. The AS individual struggles constantly to process far too much information, so determined are their brains to process every detail of what they see, hear, and feel. Thus, the familiar setting is the comfortable setting, to the extent that anything can be considered comfortable. Any new setting or change in routine represents not the suite of new information that most people would have to process in new surroundings, but an overwhelming flood of new sensory input. At the far end of the neurological spectrum, autistic individuals resort to extreme behaviors such as rocking, head-banging, or groaning, in an attempt to shut out an unbearable overload of stimuli.

Here again one finds a trait that beams out strongly as one considers Richter's experiences throughout his life. As earlier chapters make clear, he handled transitions badly and generally minimized them to the extent humanly possible. Richter was a world-class homebody throughout his life. His one early foray away from Los Angeles—his only real childhood home—ended in a nervous breakdown. Even life on a sheltered college campus had been overwhelming. Liane Willey wrote of her college years, "By the time I finished my first six years of college I was a bit beaten up, limping with failure, and deep in despair because I did not yet know why those things that came so seemingly easy to others, were so impossible for me to achieve." Willey adds, "But I was not undone. My slow descent into total confusion and overwhelming anxiety attacks did lead me to visit with a counselor on campus who gave me some of the best advice I ever received." Richter had no such counseling: in the 1920s Asperger's syndrome did not exist as a recognized disorder. The first major transition of his life led him into the same overwhelming confusion and anxiety that Willey describes, and he was undone. He returned home, and home he stayed for the rest of his life. He moved away from his family home only at the age of thirty-six, only as far as Pasadena is from Los Angeles, and only because his wife and sister were at each other's throats.

Richter did travel later in life, visiting New Zealand in 1949 and 1970 and making an extended visit to Tokyo as a Fulbright Scholar during the 1959–60 academic year. In a 1961 article in the Caltech magazine *Engineering and Science*, he described his experiences in the unfamiliar setting, one that he coped with remarkably well. He set out to learn the language, never attaining full mastery of the spoken word but learning enough of the written language to overcome the overwhelming sense of being utterly and hopelessly lost. When some in Japan demonstrated against American policies, Richter concluded that most of the Japanese demonstrators were "fairly described as liberals, not radicals, with a desire for friendly relations with Americans."

Richter's successful experience in Japan appears to belie his overall homebody tendencies, yet they illustrate a point made earlier about AS: with age and maturity the AS individual can develop skills that do not come naturally. For Richter, the baffling complexity of the written language may in fact have reduced his discomfort level, reducing the vexing social problem (to connect with people as a stranger in a strange land) to a far more comfortable intellectual challenge (to "merely" learn a language most Westerners find impenetrable).

Other transitions were harder to face, in particular those that shaped his personal life. By his own admission, for many years he found himself unhappily married, and very much in love with another woman. "There might have been—probably was—an interval when I did not really care much for Lillian, but went on with her largely from force of habit, and from unwillingness to make the effort to break loose."

Perhaps most remarkable of all are the sentiments that he expressed about seismology, twenty years after first entering the field and fifteen years after developing the scale that brought him worldwide acclaim. "After twenty years," he wrote in his letter to Dr. Moriarty, "I still have not accepted my position wholeheartedly. I have not the least intention of giving it up; I wish—and that word indicates the real lack of force behind it—I wish, and without actually doing anything about it, that I could supplement my overt occupation with something more fundamental."

Whatever else he was or wished to be, Charles Richter was an extraordinarily gifted observational scientist. Observational seismology is an applied science, one in which the problems cannot generally be solved neatly with mathematical equations or definitive laboratory experiments. Instead, observational seismology involves the business of considering large volumes of complicated data and finding answers in the midst of the chaos. Charles Richter clearly excelled at this. Colleague Robert Sharp observed, "He could look at a seismogram, size it up right away and see just where it fitted in some larger seismic picture." So, too, were his observational powers keen when he focused them on himself. The quoted letter to Dr. Moriarty reveals impressive self-awareness: that as much as he might wish to be engaged in more "fundamental" endeavors, he lacked the force of will that would have been required to act on this wish. Thus by his own account, even when he yearned deeply to make changes in his life, he lacked the wherewithal to follow through. As a result, Richter spent nearly his entire life within a markedly limited turning radius.

General Anxiety, Outbursts of Temper

As Bashe and Kirby discuss—and Richter himself described in verse—life is a never-ending stream of assaults on the senses for individuals with Asperger's syndrome. With every social interaction and virtually every task fraught with peril, people with AS live their lives under a much higher level of ambient stress than do most normally social, functional human beings, for whom day-to-day life is simply easier. The AS individual may thus experience a pervasive sense of anxiety on a nearly constant basis, and be less able than most to roll with the punches in the face of minor provocations.

Anxieties shine brightly through much of Richter's writing, perhaps none more so than the graduate school journal, in which he anguished over everything and anything. The outbursts of temper, meanwhile, were among the facets of his complex personality that have been well known among the seismology community, and certainly among his colleagues at Caltech. "He could also be explosive," colleague Francis Lehner said, "and really blow up over something." Yet at his core, Richter was not fiercely temperamental. "But whenever he lost his temper like that," Lehner continued, "he'd feel really bad afterward and try to make it up to you."

Colleague Eric Lindvall, with whom Richter launched a consulting business following his retirement, echoed Lehner's observations. Lindvall recalled hearing tales of staff meetings at Caltech, meetings during which something or someone stoked Richter's ire. On one occasion he stormed out of the room, slamming the door so hard that its pane of glass shattered. After a few seconds of dead silence in the room Richter returned to inform his colleagues, "I am not sorry about what I said but I am sorry about the door." On a second occasion he fumed and stomped out of the room, grabbing Gutenberg's hat off the rack in his haste. He again returned momentarily, slamming the hat onto the table so forcefully that he punched a hole through the top of the hat. Richter did not say anything at the time but a couple of days later a new hat, neatly boxed, appeared on Gutenberg's desk.

Tic Disorders

The best known of the so-called tic disorders is Tourette's syndrome, a neurological disorder characterized by motor tics, involuntary movements, or vocalizations. Symptoms range from, at the worst, severe enough to interfere with an individual's ability to function in society, to, at the more mild end of the spectrum, simple but repeated movements like eye blinking or throat clearing. Like AS, Tourette's occurs more frequently in boys than in girls. Tourette's is considered a separate neurological condition from Asperger's; the two by no means always occur together. Other neurological conditions can arise with AS as well: if one thinks of the AS brain as being somehow incorrectly wired, other sorts of wiring problems might exist as well. These other conditions include obsessive-compulsive disorder, seizure disorders, attention-deficit-hyperactivity disorder, bipolar disorder, and more. Richter was by no means afflicted with all of these; a tic disorder, however, is nearly certain. Those who knew him describe a pronounced facial tic, present much although not all of the time: persistent, jerky blinking of his eyes. Such a disorder would only have served to complicate the already difficult social interactions of an AS individual.

Asperger's Syndrome: The Full Package

Life is neither kind nor easy to children unlucky enough to be born with Asperger's syndrome. They often behave in ways that frustrate parents enormously, seeming for all the world not to listen to what they are told or to care what their caregivers or peers think or feel. They may act in ways that seem deliberately obstinate or even hurtful. An AS child might well impress caregivers as a juvenile delinquent in the making, if not one already fully made. Without a correct diagnosis, the AS child might be labeled ADHD, or thought to have an oppositional disorder. In the absence of any diagnosis, for example in earlier times or in families who lack the means to seek professional help, the AS child might simply be considered a pain the ass. One wonders how many undiagnosed AS children are physically abused by well-intentioned parents who find themselves at wits' end with a child who simply will not listen, will not behave.

Richter was less disruptive or defiant than some AS sufferers due to his profoundly introverted temperament. Limited social skills would be far less bothersome to an individual who was inclined to live in his own head in any case. (A tendency towards introspection is one of the reasons that girls with Asperger's syndrome might go undiagnosed.)

Yet even if Richter did not impress adults as a juvenile delinquent as a child, his adult behaviors and actions were clearly seriously and repeatedly misunderstood by those around him. His inability to laugh at himself, or at least at certain kinds of jokes, did not, as some assumed, reflect an absence of a sense of humor. Those who knew him well came to appreciate his dry wit. Richter's inability to bring professional projects to completion, meanwhile, did not, as Hertha Gutenberg may have assumed, indicate laziness or a simple lack of willpower. By all accounts Beno Gutenberg's own mind was extraordinarily wired for any number of things, organizational skills included. Richter's mind, although brilliant in its own right, was as inherently disorganized as Gutenberg's was organized. Someone in the latter position would almost invariably fail to understand what the former position is like.

Similarly, Richter's apparently diffident interactions with humanity emerge as a function not of hostility, but rather of unease. Simply put, people were difficult for Richter to deal with. David Johnson observed that "Richter either liked you or he didn't," and "didn't have much use" for people with whom he did not get along. Viewed from the AS perspective, one suspects that some people were not so much especially irksome to

Richter as they were especially alien. One further suspects that a certain breed of student, for example, would have left him especially flummoxed, in particular students and scientists who were more assertive and social than your typical introverted scientist. Certainly classrooms full of students left him flummoxed: he was known for delivering entire lectures speaking directly to the blackboard, never turning around to face his students.

Animals were another matter: a safer target of affections than people. The Richter household was rarely without a cat: there was Skunky (who, as Richter instructed occasional pet sitters, was "crazy about cheese"), then Temblor, then Shomyo kitty, then Oliver. In a 1965 letter written to Lillian during her extended trip abroad, Richter wrote, "[The cat] seems pleased; we have had a little argument lately, because the heart I got on Sat. has too much of that white lard with it, and he disapproves even when there is better meat in the dish; fusscat. So this evening I have separated out mainly the cleaner pieces for him." He went on to say, "I should have been much lonelier and less cheerful without kitty; he is a sweet purry thing." If one can read his novel draft, "House on a Bridge," as autobiographical, Richter might have preferred dogs to cats, as did the character who emerges as Richter's alter ego. Indeed, later in life, after Lillian's death, Richter did own a dog named Jack. But Kate Holliday, who clearly spent a considerable amount of time with Richter, described him as a "linguist, devoted amateur botanist, cat-lover, [and] wit." She also wrote about his "greatest devotion" next to his wife, the "thick-haired beast [who] obviously regards Richter as a great patsy and demands endless hours of petting, scratching, brushing, and other forms of attention. Richter not only complies, but has filled his bathroom with color shots of the animal and has tucked small ceramics of felines into odd nooks about the house." Longtime Seismo Lab technician Bob Taylor also recalls Richter bringing in cat food to feed the semi-feral cats who made their home in the hills around the Kresge Lab.

Richter was not, however, disinterested in human friendships. He positively reveled, in the nude, no less, in his first discovery—in his mid-thirties—of truly intimate friendship. In 1945 he wrote candidly of his "willingness to give everything from myself that I might received [*sic*] the most from others." He was, as we have seen, absolutely not uninterested in women: as a young man as well as an older one he clearly very much craved physical as well as emotional affections of members of the opposite sex. He celebrated the few close women in his life as individuals. Recall his own inimitably clumsy, but fundamentally appreciative words, "Only a few have loved me, but those few, / Are not more like by night than like by day."

That Richter's love life was rather more complicated than most—that those "few" may have included his own sister—can perhaps be better understood as well. Sexuality can be difficult for any AS individual. Quoting again from the OASIS Guide, "The lack of social skills can place teens and adults with AS at risk because they may not understand the intentions of others or be able to express their own intentions clearly." Poetry, moreover, provided not only a key creative outlet for Richter; it provided the one medium by which he connected with other people. He shared his poetry, and bared his soul, to fellow writers—to women—whom he knew as part of his life as a writer; some of these friends became lovers. As for the possible relationship with Margaret, the likes of which humans are strongly programmed to avoid, one can return to the point that Richter's makeup appears to have differed substantially from the norm; this would have affected many if not all facets of his life. At a minimum this much is true: Richter lived a life largely apart from usual societal conventions.

A posthumous diagnosis of Asperger's syndrome of course remains speculative, the abundance of direct and circumstantial evidence notwithstanding. Nor would this diagnosis change the fact that Richter was, by societal standards, a freak or a geek—a man for whom social interactions were vexingly alien. However, this diagnosis—if we accept it—provides a framework from which apparently negative, sometimes infuriating or disheartening personality traits emerge in very different light. Richter was tormented throughout his life by demons of the most persistent, insidious, deeply rooted kind. By virtue of nonstandard brain wiring he found himself a Martian on a planet of Earthlings, struggling throughout his life to relate to these inscrutable creatures and master the subtle points of their language. He struggled to maintain the sort of life that he felt a man was supposed to lead: a life of easy confidence, of social contacts, of accomplishment, of tackling jobs and tasks with appropriate organizational wherewithal. He struggled to focus his boundless energies and prodigious intellect in productive directions rather than allowing his demons to carry him away. In this last battle he succeeded admirably, but by no means easily and certainly not unfailingly. His monumental professional achievements thus emerge as all the more impressive—and his personal tribulations, all the more sympathetic.

CHAPTER 15

Here It Comes Again

From my office door I could see the ink-writing instrument begin to deflect, just as I felt an earthquake shaking the building.
—*Charles Richter, "Here it comes again," 1971*

ANY seismologist's research and life will be profoundly shaped by the earthquakes that strike during his or her career. In Richter's case two events were arguably of paramount importance, a pair of seismic bookends that not only bounded but also gave definition to the bulk of the career in between: the 1933 Long Beach earthquake and the 1971 San Fernando, or Sylmar, earthquake. Ironically neither of these was the largest temblor that struck Southern California during Richter's career; that distinction goes to the magnitude 7.5 earthquake that struck before dawn on July 21, 1952, to the north of Los Angeles. The so-called Kern County earthquake caused fifty million dollars in property damage and claimed twelve lives. Damage and loss of life were most severe in the central California community of Tehachapi. People in Reno, Nevada, as well as some individuals on the upper floors of tall buildings in San Francisco, felt the ground or building sway beneath their feet. Shaking was strong enough in Los Angeles to cause some damage and power outages. We have already heard about the immediate effect of this earthquake on Richter's life: thirty seconds of swearing brought upon by the realization that his vacation plans had been ruined. The earthquake would have a substantial longer-term impact on his life as well. He once again collected portable seismometers and headed out to the field to record aftershocks, of which there were many.

The Long Beach earthquake had been followed by a large number of aftershocks, one of them large enough to cause additional damage. The aftershock sequence following the Kern County shock put this to shame.

A magnitude 6.4 aftershock struck about thirty minutes after the mainshock, and a magnitude 6.1 event occurred two days later. It might be useful at this juncture to highlight a sometimes overlooked fact about aftershocks: they are earthquakes. A magnitude 6.4 earthquake is for all intents and purposes, most notably the shaking it generates, the same whether it is a mainshock or an aftershock. Aftershocks can thus be more than a little unruly: they can be lethal.

The Kern County mainshock and the two early large aftershocks were fortunately centered in a sparsely populated part of south central California. Had the quakes struck in more urbanized area, the property damage and loss of life would have been very much worse. On August 22, a relatively modest aftershock did strike closer to a population center, a magnitude 5.8 shock centered near the town of Bakersfield. This moderate earthquake caused two deaths and an additional ten million dollars in property damage, mostly to brick buildings in the downtown area.

At magnitude 7.5, the mainshock itself was a portentous event: shocks of this magnitude occur in California only once every few decades on average. Data from portable seismometers allowed Richter and his colleagues to determine aftershock locations with greater precision than would have been possible with data from permanent network stations alone. Aftershock locations as a rule illuminate the extent of the fault that moved during a mainshock: in this case a patch of the White Wolf Fault. The earthquake did not break the surface cleanly over its extent but it did leave a jagged, disjointed surface rupture that in some locations involved substantial movement across the fault. This movement cut directly across tunnels on the Tehachapi Pass railroad route, putting the rail line out of service for three weeks.

When an earthquake such as the 1952 temblor occurs in a relatively farflung corner of Southern California, its impact on seismologists far outstrips its impact on most people in the greater Los Angeles area. A similar story played out almost exactly forty years later when the magnitude 7.3 Landers earthquake tore through an even more remote desert region in 1992. For the majority of Southern California residents Landers was very nearly a nonevent; for seismologists it was life-changing.

Landers proved to be a watershed earthquake for seismologists, for several reasons. Among these was the first ever (or so we thought) recognition that large earthquakes can be followed by not only aftershocks but also smaller quakes that were dubbed *remotely triggered earthquakes.* These triggered earthquakes are like aftershocks in the sense that they are caused by the larger mainshock, but they occur at much greater distances from the mainshock. Research done in the 1990s provides compelling evidence that they are triggered by the waves that ripple through the crust when a large earthquake strikes. The recognition of remotely triggered earthquakes has been part of an important paradigm shift in seismology in the years since 1992: whereas seismologists once saw most earthquakes as random and disconnected, since 1992 we have recognized a much greater degree of interaction—of *communication*—between some of them.

In 1994, seismologist Joan Gomberg, who did some of the pioneering research on triggered earthquakes, wrote that "no previous experience would have led us to anticipate the observation of remotely triggered earthquakes." Nobody begged to differ at the time, but, as it happens, she was wrong. In May 1955 Richter typed what apparently remained unpublished notes: "The reverse effect—a major earthquake triggering minor shocks—is most probably [*sic*] within the immediate aftershock are [*sic*], but essentially by elastic wave propagation it may set off action at a greater distance." He went on to write, "If the (triggered earthquake) is large enough, it may itself act as a trigger, so that there may be relay action, in which some of the later events are larger." Richter thus not only recognized the remotely triggered earthquakes that were first observed by other seismologists in 1992: he had the correct explanation for them! One wonders which earthquake sequences had led him to this understanding: perhaps it was the Kern County earthquake itself, which struck near Tehachapi and was followed about five hours later by a small burst of earthquakes near Riverside, over one hundred miles away. One also wonders how much more Richter might have figured out, had he had at his disposal the volumes of data that would be collected for later earthquakes such as Landers. In any case, the Kern County earthquake did not spark the kind of scientific excitement that Landers did forty years later. It might have changed seismologists' thinking, and Richter's career—had he had better data, or had he had the wherewithal to pursue and publish his observations and ideas. But he didn't; because of this, and the fact that the Kern County earthquake struck a relatively remote area, the biggest earthquake of Richter's career did not become the most important earthquake of his career.

Charles Richter was not around to witness the Landers earthquake, and the revolution it inspired, having died seven years earlier at the age of eighty-five. However, his long career was punctuated at the end by the temblor that broke a decades-long earthquake drought in the metropolitan Los Angeles area: the 1971 San Fernando event, which struck the northern San Fernando Valley on the morning of February 9. (Earthquakes did of course occur in the Los Angeles area between 1933 and 1971, but these decades witnessed a marked paucity of temblors large enough to cause even light damage.) Seismologists generally know this as the Sylmar earthquake, the town within the San Fernando Valley closest to the epicenter.

By 1971 Richter had retired from Caltech, although he remained active in seismology circles and by that time had a seismometer operating in his living room. The initial shaking was strong enough and long enough to tell him that this could be the beginning of a major earthquake. In Richter's words, however, "after thirty seconds the perceptible shaking, and the recorded oscillations on the seismogram, were not increasing, so it was clear that we had to do with another moderately large event, perhaps a local disaster." The words "moderately large" might sound strange to anyone who experienced the worst of the temblor's wrath. For anyone in or close to the San Fernando Valley on that February morning, the earthquake could scarcely be considered only "moderately large." Yet even from the initial shaking and recording, Richter recognized the temblor to be a relatively modest event—its magnitude later estimated at 6.6–6.7–rather than a truly major earthquake of magnitude 7.5 or higher. During those initial thirty seconds when he wondered whether or not the shaking would build, one suspects Richter wondered if the proverbial Big One had struck the San Andreas Fault at long last.

For those in communities like Sylmar and San Fernando, however, the Big One *had* struck. The temblor caused a half billion dollars in property damage and claimed sixty-five lives. It highlighted in dramatic fashion the vagaries of California's approach to building codes: although the Field Act had guaranteed a substantial measure of safety for public school buildings, no similar law had been passed for the state's hospitals. Most of the deaths occurred in older but well-built buildings of the Veteran's Administration Hospital near Sylmar. Two of the older buildings collapsed, killing more than forty people. The relatively new Olive View Community Hospital in Sylmar was a total loss. Once again the timing of the temblor proved fortuitous: the earthquake struck in the early morning hours, when most area residents were asleep in their homes. California homes are generally of wood-frame construction and have for decades been built to meet building codes that include earthquake design provisions. As a consequence, Californians are generally safer at home during an earthquake than they would be at work or on the freeways.

Shaking would have felt severe enough (to most people) in Pasadena, but would have caused little damage to a well-built single-family home. Richter thus had no personal disaster on his hands but understood the

Fig. 15.1. Damage to single-family house caused by 1971 San Fernando earthquake. (Photo courtesy of National Geophysical Data Center.)

extent of the impact on the operations of the Seismo Lab and his colleagues' lives. He therefore got dressed and prepared to go to the lab, although "not in too much haste." Since his retirement he was no longer familiar with the detailed daily operations of the lab, and knew that others could handle the initial data analysis better than he could. By the fall of 1971 Lillian's health had also begun to fail, a consequence of her ongoing battle with colitis, and possibly a consequence of the cancer that would eventually claim her life. One wonders if she didn't weigh heavily on his mind as well.

Richter ate a quick breakfast and fielded a few telephone calls from home, although, "not to much purpose, since I could give no details, and hesitated to offer estimates which might cause confusion when better information became available." Listening to reports on the television and radio, he learned that the main NBC studio in Burbank had lost power and was broadcasting from a portable unit in their parking lot. He left for the laboratory at about seven-thirty, an hour and a half after the earthquake had struck. (As a point of reference, most seismologists pause just about long enough to put clothes on before dashing off to the lab after a major earthquake strikes close to home.)

Although he had remained involved in seismological matters following his retirement, by 1971 Richter was removed from the fray that largely swirled around him. Teams of scientists and technicians left the lab to deploy portable seismometers, far more numerous and sophisticated than the portable devices of Richter's day. The epicenter, or precise point of origin, of the earthquake had been pinpointed by the time Richter arrived at the lab.

Richter learned through the media that the Lower Van Norman dam had nearly failed. Having earlier served as a consultant to the Department of Water and Power, Richter had been aware of safety issues concerning the dam and convinced DWP to lower the reservoir before the earthquake struck. In fact, the dam was badly damaged. The structure sustained considerable damage, and in Clarence Allen's words, "We missed a major catastrophe by only a whisker": the main portion of the dam had held, barely, thanks in part to the lowered water level. Residents below the dam were evacuated for the two to three days that it took to bring the reservoir level down further. Former student John Gardner recalled the DWP's fear that a large aftershock would cause the damaged dam to fail catastrophically. On average, the magnitude of the largest aftershock will be about one unit smaller than the mainshock. But as with almost everything in matters seismological, the earth is not bound by a law of averages: in this case, fortunately for those living below the dam, the anticipated magnitude 6-ish aftershock never happened.

In 1971 as now, the large temblor generated a prodigious response from local news media, members of which soon crowded the laboratory lobby. Only when the same scene played out to an even greater extent following the 1994 Northridge earthquake did Caltech build (with a grant from the Times Mirror Foundation) a well-equipped media center for such occasions. The scene in 1971 was mostly one of mayhem. Richter talked with reporters, by his later account, "to take some of the pressure for information and comment off of the working staff." By this time he was clearly sensitive to the notion that he actively sought out press attention.

As is often the case, the crush of media attention had a few unfortunate consequences as scientists rushed to get reporters the best possible information and then, as necessary, to issue updates. The initial magnitude estimate of the earthquake had been determined not by scientists in Pasadena but by seismologists analyzing data at the University of California at Berkeley as well as at the national earthquake center in Washington, D.C. An earthquake as large as San Fernando generates waves that can be recorded all over the world by seismometers designed specifically to record

Fig. 15.2. Damage to Van Norman Dam caused by 1971 San Fernando earthquake. (Photo courtesy of National Geophysical Data Center.)

large, distant earthquakes. Data from such instruments were by 1971 used routinely to obtain reliable magnitude estimates, and yielded an initial estimate of 6.5 for the San Fernando temblor. The laboratory staff at Caltech would have trusted their colleagues' results: thus was the magnitude estimate released to the press and public. Efforts in the Pasadena lab remained focused on determining accurate aftershock locations, as these were critical to guide the deployments of portable seismometers. Indeed, early results revealed a belt of aftershocks extending beneath the San Fernando Valley to the west of where the mainshock rupture had first appeared to be, leading to a redirection of field efforts. Seismo Lab staff did not calculate a magnitude from their local data until the next day. When they did, their answer—which refined the earlier solution using better data—was higher than the preliminary value. Eventually the preferred magnitude estimate settled at 6.7.

Considering the limitations that existed at that time in rapid magnitude determinations, the preliminary 6.5 value had come admirably close to the mark. Yet the initial reported value was subsequently revised, leading to, as Richter put it, the "myth of the 'broken' magnitude scale." Nearly forty years after his seminal publication describing the Richter scale, many still had the mistaken impression of the scale as a mechanical device.

In retirement Richter remained active in seismology as a consultant and in a variety of other capacities, including the enduring role of earthquake luminary. His active participation in research had, however, by and large come to an end. Thus was his role in the aftermath of the 1971 temblor largely one of elder statesman, consultant, and spokesman. In fact the earthquake led directly to a larger postretirement role than Richter might otherwise have had: it provided the impetus for Eric Lindvall to approach Richter with a proposal that they join forces to form Lindvall, Richter and Associates. Lindvall's father had earned his Ph.D. at Caltech in the 1920s, although he had not known Richter at the time. Eric Lindvall had by 1971 earned a masters degree in geology from Stanford University.

Lindvall, Richter and Associates would go on to consult for a number of key clients over the following decade, including government agencies. Immediate after the 1971 earthquake Richter served as a consultant in an individual capacity, especially with agencies such as the Department of Water and Power, for whom he had done some consulting prior to his retirement. The DWP sought him out the day following the earthquake to determine how long the evacuation should be maintained below the damaged Van Norman dam. In this capacity he reminded officials of the ever-present hazard of damaging aftershocks that might well damage an already weakened dam.

Richter's advice to the DWP found its way to the news media in unfortunately corrupted form: some broadcasters reported that Charles Richter had confidently predicted that another shock "almost as big" as the mainshock would strike. As he later said, "Quotations were further confused by a device employed by journalists and broadcasters on every such opportunity. Seismologists are then invariably questioned about the possibility of a great quake in the area." Richter further noted that it was a reasonable question to ask: had the 1971 temblor postponed or hastened the great earthquake that scientists know will occur some day on the San Andreas Fault to the north of Los Angeles? "To this," he noted, "the proper answer is that we do not know. My guess is that even earthquakes of magnitude 6.5 do not much affect the steady accumulation of strain toward the major break, which will come in its own good time."

Had Richter's career ended a couple of decades later, he would have witnessed developments that now allow seismologists to respond to such questions at rather more length. Scientists still cannot answer the above question with certainty, but they can now calculate the effect that an earthquake such as San Fernando will have on neighboring faults such as the San Andreas. Richter's highly educated guess was a good guess: a temblor of magnitude 6.5 will not significantly affect a fault as far away as the San Andreas Fault is from San Fernando.

However, other recent developments have suggested a somewhat different take on the so-called earthquake cycle: the process whereby strain builds towards the "major break," or great earthquake, and then is released. As Richter astutely surmised, if moderate earthquakes are the tail, then great earthquakes are the dog, and the tail does not wag the dog. Which is to say, the largest earthquakes in any area are the fundamental events, the driving events: they control the timing of everything else, not vice versa. But how, exactly, does the dog wag the tail? In the closing years of the twentieth century, geophysicists Geoffrey King, David Bowman, and Charlie Sammis proposed a new take on the business of dogs and tails. In their model for the earthquake cycle, strain energy accumulates in the crust as the major fault builds towards a great earthquake; this strain can generate a spattering of moderate shocks in the neighborhood of where the future great earthquake will occur.

In fact Richter had also anticipated this idea. Following the 1933 Long Beach earthquake he wrote a letter to Caltech head Robert Millikan outlining his concerns for possible future earthquakes: "The more reassuring view holds that the minor destructive shocks act as a sort of safety valve, to mitigate or even prevent the accumulation of strain which would otherwise be relieved in a major earthquake. The opposing view, though admitting there must be some such mitigating effect, holds that it is of minor importance, and that the minor destructive shocks are to be looked upon primarily as relieving minor and superficial strains incidental to the accumulation of the major strain, so that they are warnings of increased danger rather than assurances of safety." Richter went on to discuss the "historical evidence bearing on this question," and pointed to the observation that between 1890 and 1906, there were five or six moderate shocks in the "province visited by the major shock of the latter year."

In 1999, sixty-six years after Richter's letter to Millikan, geophysicist Ross Stein published a paper in *Nature* that discussed the earthquakes in the San Francisco Bay Area in the decades before versus the decades after the 1906 earthquake. His analysis confirmed and extended Richter's basic observation, and yet one is inclined to be impressed at Richter's degree of prescience at such an early date. The depth of his intuition reflected an enduring reality in earthquake science: the most important new ideas often spring directly from new data that illuminate a hitherto unrecognized phenomenon. Through the better part of the twentieth century, no scientist was closer to the data—or, arguably, more able to unlock their secrets—than was Charles Richter.

The model of King and his colleagues provides a physical explanation for an observation that moderate shocks do often seem to strike with increasing frequency in the area surrounding a future large earthquake. In effect, the dog has tremendous pent-up energy that it cannot immediately release by shaking all over, and so the tail wags. From this viewpoint, then, a temblor such as San Fernando does not hasten the occurrence of a future great earthquake, but rather might be a harbinger that a great earthquake is on the way. Such an inference, however, cannot be based on the occurrence of a single moderate shock, but must be made based on patterns of moderate earthquakes in an area over a span of a few years to a few decades.

Seismologists' understanding of the earthquake cycle remains rudimentary at best; some respected scientists have different ideas about dogs and tails. The new developments in the field do, however, provide the basis for somewhat more thoughtful discussion than the only one available to Richter in 1971: "We don't know." Although in the final analysis, the bottom line remains the same.

Richter observed that certain types of questions ("Has this temblor hastened or postponed a future great earthquake?") provide the "opportunity for distortion, since it is easy to confuse the constant indefinite expectation of a large earthquake with the immediate and comparatively certain expectation of aftershocks." Recent developments in the field notwithstanding, this observation also remains as true today as in 1971.

From his vantage point, removed from the day-to-day operations of the Seismo Lab, the San Fernando earthquake likely provided Richter with the best opportunity of his life to consider the media response to a major earthquake. The day the earthquake occurred he returned home later that same morning, catching the live television broadcast of the *Apollo 14* splashdown. "In the intervals," he wrote, "I found several channels giving good coverage of the earthquake." He credited a number of local channels for their interesting and informative reports, in some cases including helicopter footage of the reservoirs and most heavily impacted regions. He did, understandably, take exception with one unnamed broadcaster who "tried to enliven his program by introducing a burlesque interview with a soothsayer who was represented as prophesying another catastrophic earthquake." This broadcaster later expressed surprise that anyone might have taken such an obvious farce seriously. "Even a 'disc jockey,'" Richter noted with palpable irritation, "should know that a time of disaster is not a proper occasion for tomfoolery."

Every major earthquake represents a tremendous learning opportunity for seismology, so few and far between are the events that provide seismologists with their most critical data. Richter understood that even the relatively modest San Fernando temblor would have much to teach seismologists about the nature of faults and earthquake hazard in the greater Los Angeles region. He wrote of the potential for large earthquakes on other faults in the region, including one that would produce the damaging magnitude 5.9 Whittier earthquake in 1987, two years after his death. He wrote about the Raymond Fault, which cuts in an east-west orientation across the San Gabriel Valley, passing through the city of San Marino just a mile or two south of the Seismo Lab. Former DWP engineer Le Val Lund echoes Lindvall's admiration of Richters' work as a consultant, which involved seismic evaluation of dozens of DWP dams. Lund describes Richter as having been congenial and well-liked by the staff.

One can only wonder what thoughts ran through Richter's mind in the aftermath of the San Fernando earthquake—the biggest earthquake disaster in the greater Los Angeles region in his lifetime. Perhaps a measure of regret, to be out of the fray in the midst of the excitement; perhaps a measure of relief. Or perhaps his attentions were merely focused elsewhere at the time—at home, where his wife's health continued to decline. In any case the earthquake, the closing seismic bookend of Richter's long and distinguished career, gave him one last moment in the sun, the chance to step into the elder statesman role. As an earthquake ambassador, it fell to him to serve as liaison between the scientific community and the media and public: to answer questions, offer advice to officials, and dispel myths as best he could.

Charles Richter became a minor celebrity to the residents of Southern California in the days and weeks following the 1971 earthquake. By this time the television era was well under way: Richter thus became not only a famous name but also to some extent a television star. The earthquake also effectively brought him back into a more substantial role as a consultant, one in which he would continue to serve for another decade. When author Digby Diehl set out to interview celebrities for his 1974 book, *Super-Talk*, he picked Richter—along with more conventional media stars like Joan Baez, Gloria Steinem, Huey Newton, Ted Geisel, Norman Lear, and Hugh Hefner. Whereas the 1933 Long Beach earthquake had cemented his commitment to the citizens of Southern California, the 1971 San Fernando temblor in turn cemented the citizenry's commitment to Charles Richter.

CHAPTER 16

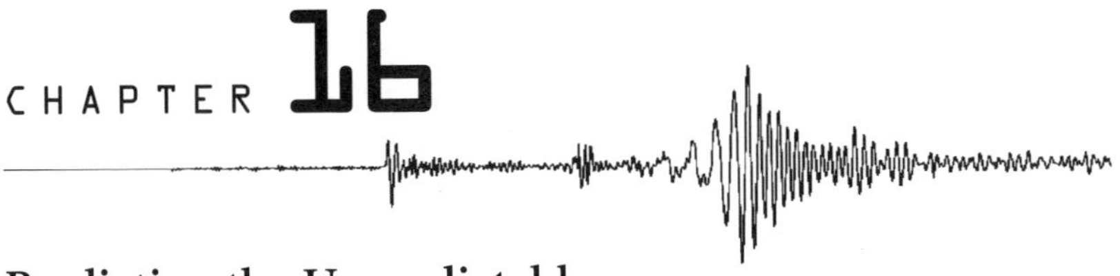

Predicting the Unpredictable

> Only fools and charlatans predict earthquakes.
> —*Charles Richter*

Ernest Rutherford, Nobel Prize winner in chemistry in 1908 for groundbreaking research on radioactivity, observed famously that "Science is physics and stamp collecting." While this assessment might sound like an affront to scientists of many persuasions, most scientists have an appreciation for the sentiment: Modern science is expected to go beyond simple observations and descriptions of phenomena ("stamp collecting") and explore the reasons why things are the way they are ("physics"). If there is a single litmus test for modern science, it is whether our understanding provides not only a description of past events but also the basis for prediction of the future.

Yet, today as in the 1970s, prediction remains a source of frustration to seismologists. Our inability to predict earthquakes prevents us from mitigating disasters such as the world witnessed with horror and sadness on the day after Christmas of 2004, when a powerful earthquake in the Indian Ocean near Sumatra caused a massive tsunami that left hundreds of thousands dead and changed the very shape of many islands and coastlines. For all of the money spent on earthquake research, for all of the recent developments in understanding, people sometimes wonder—and not altogether unfairly—what good are seismologists if they can't predict earthquakes?

Seismology is scarcely, however, without predictive capability. We can, for example, predict with a fair measure of certainty how the ground will shake during an earthquake on a given fault. Based on work dating back to Gutenberg and Richter's 1944 study (and Ishimoto and Iida's 1939 publication), we can also predict with a fair measure of certainty how many earthquakes of various magnitudes will strike a given area *on average* in a given time period. But earthquake prediction per se remains elusive, always seeming to lurk just around the next corner, almost but not quite within reach. The term *prediction* has a very specific meaning, one that is generally well understood in both public and scientific circles: a valid earthquake prediction is one that specifies, within useful narrow windows, the precise time, location, and magnitude of a future earthquake. In fact, however, this definition is not as clear-cut as it seems. Among the games that charlatans and fools play is to take liberties with the phrase, "useful narrow windows"—liberties that, while subtle, can make an unsuccessful or useless prediction method look successful upon casual inspection. But I will return to that part of the story later.

The inability of scientists to predict earthquakes generates understandable frustration among the public. There is little in nature so fundamentally terrifying as a large earthquake commencing with zero warning, out of a clear blue sky—or worse, a jet black one. Human notions of stability are inexorably intertwined with the stability of the ground beneath our feet, as revealed by familiar phrases such as *rock solid, terra firma, well grounded.* It challenges our sense of stability and security, at a very fundamental human level, when the terra rudely and catastrophically ceases to be firm. The unpredictability of earthquakes also, of course, accounts for their sometimes lethal nature: if you don't see them coming, you can't get out of the way.

Seismologists are not uninterested in earthquake prediction; research efforts have continued, at times quietly, as scientists continue what has been termed a quest for the Holy Grail. (One should note that earthquake prediction is scarcely of paramount concern for all seismologists, many of whose research focuses on the structure of the earth.) Respected scientists have in recent years proposed prediction methods that, if not able to pinpoint the precise date of a future earthquake, at least seek to predict within a few months or years when a large earthquake is likely to strike. Such efforts are fraught with peril, however, by virtue of the media attention they are likely to garner. In contrast to economics, where predictions often seem to be right and wrong in equal measure, people look to earthquake scientists to take away the harrowing element of unpredictability, and tend to apply a zero-tolerance standard to any emerging earthquake prediction efforts. People want earthquake prediction; they don't want false alarms.

Within the United States the popularity, or perhaps the acceptability, of earthquake prediction research has waxed and waned over the years. Richter's formal 1979 interview with a Caltech archivist provided him with an opportunity to look back at half a century of prediction-related research. This interview represents a retrospective of prediction research in the United States during the latter half of the twentieth century. By 1980 prediction research had fallen into disfavor among the U.S. seismology community and apart from the Parkfield Prediction experiment, little work was done in the closing decades of the century. As Richter recounted, the 1964 Alaskan earthquake—a monstrous magnitude 9.2 event—had gotten the ball rolling in the first place, generating a "tremendous clamor for earthquake prediction." "And this," he added, "I regret to state, was in part aided and abetted by some of my own colleagues [at Caltech]."

Richter's comments referred at least in part to the case of the Palmdale Bulge, which sparked a firestorm in public and scientific circles in the 1970s. As I will discuss shortly, the furor began with what proved to be a flawed study by a scientist employed by the U.S. Geological Survey. To set the stage, however, it is useful to consider an earlier earthquake prediction dustup in California, one in which the roles were rather more complicated, and different from, those in the later incident. In the mid-1920s, an eminent geologist named Bailey Willis set off a firestorm with dramatic, alarmist statements about future earthquake hazard in Southern California. He argued that a future great shock, on a par with the great 1906 earthquake, might strike Southern California sooner or later, but claimed it was more likely to be sooner. By the mid-1920s Willis had retired from an earlier position with the U.S. Geological Survey and a professorship at Stanford. His "prophecies," as Robert T. Hill described them in his benignly named book *Southern California Geology and Los Angeles Earthquakes*, had been based on analysis of early surveying ("triangulation") data published in 1924 by the Coast and Geodetic Survey. His analysis purportedly showed that the two sides of the San Andreas Fault had been trying to slide past each other at the astonishing rate of twenty-four feet in thirty years. Because the fault itself is locked and cannot actually slide smoothly, the result implied that an enormous amount of strain, or energy, had been building on the southern San Andreas in very recent years. In fact, modern data show that strain is building on the southern San Andreas Fault, but at a rate of about three feet in thirty years.

To Willis, a geologist but not a seismologist, the implication appeared inescapable: the fault was building towards a great earthquake, probably very soon. Willis put his prognostications in print and went in front of both the Board of Fire Underwriters and the National Board of Underwriters in New York. As an immediate consequence, earthquake insurance rates in

Robert Hill's book represents an interesting milestone: in it, he described for the first time many of the now-recognized faults in the greater Los Angeles metropolitan area, but he went on to dismiss all of them as inactive, "largely things of past geologic time." Of the Inglewood Fault specifically he wrote, "it cannot be said that there is any great menace." The 1933 Long Beach earthquake struck on this fault just five years later. Hill had a distinguished scientific background, yet after his book was published, some colleagues suspected that he had sold his soul to the Los Angeles Chamber of Commerce—which rushed to embrace the book.

Los Angeles soared by anywhere from 100 percent to 2,200 percent of previous rates. In Hill's estimation, "The cost of this advance in earthquake insurance rates has probably exceeded in dollars and cents many times the total losses from earthquakes in California within the period of human history or those apt to be incurred for hundreds of years." Hill was inclined towards hyperbole in this statement (see sidebar above), but nobody can dispute the unfortunate consequences of Willis's approach. The results that he had relied on to make his predictions were badly flawed, as the acting director of the Coast and Geodetic Survey knew at least as early as 1927.

In the intervening years, however, the damage had been done. "The author of these fearsome predictions," Hill wrote, "is a scientific man of standing but in my opinion he has, in this instance, done more to discredit science in the minds of people of this vicinity than any other incident in recent years." The later Palmdale Bulge incident featured different particulars and different actors and agencies cast in the villain role, yet the two incidents reveal striking parallels. But there were other more minor crises in the decades following the Willis fiasco, including one that again may have been fueled by Willis. A March 13, 1933, International News Service story out of Stanford reported that at a meeting of insurance company presidents in New York City, "Dr. Willis [had] predicted the recent earthquake." The article quoted Willis directly: "At that time I said the earthquake might be expected within 3, 7, or 10 years." One can never be certain that any given newspaper article reflects accurately a scientist's own words or position, but the article gives one the impression that Willis did not go to great lengths to dispel the notion that his prediction was in any way fulfilled by the 1933 earthquake. (Not only had Willis's prediction been based on entirely faulty data; it had also been for an entirely different fault.)

A letter from the early days of Richter's career, dated March 20, 1933, reveals a somewhat surprising measure of alarmism from the man who would later make disparaging remarks about his colleagues' alarmism. The letter, parts of which we have discussed in an earlier chapter, was written to Robert Millikan after the 1933 Long Beach earthquake. It outlined Richter's concerns that the region, "may at any instant be exposed to an earthquake having an energy of the order of *a thousand times* that of the Long Beach shock." He went on to justify this conclusion with a consideration of prior large and great earthquakes in California and Nevada, in particular his view that "minor destructive shocks are to be looked upon principally as relieving minor and superficial strains incidental to the accumulation of major strain, so that they are warnings of increased danger rather than assurances of safety."

Richter further admitted, "Certainly any positive conclusion that the major earthquake danger in Southern California has been postponed is rather unwarranted. There is a small chance that such is the case; but it is far more likely that the danger is immediate."

There is, however, a world of difference between relaying an alarmist (albeit educated) view to a colleague and relaying it to public officials or the media. One can understand how a scientist could take the former step himself and still feel a sense of outrage with a colleague who takes the latter leap. (Richter's own colleague, Harry Wood, had in fact taken the latter leap, making unfortunate statements to the media in the aftermath of the Long Beach earthquake.) Still, where earthquake prediction is concerned, wisdom tends to come with advancing years. The young seismologist watches patterns develop that appear alarming based on past experience or existing theories; more often than not, the fears prove to be groundless, and the older seismologist emerges as so much the wiser.

Nonetheless, during the 1970s, as Richter's own career was waning, the U.S. earthquake prediction program was to a large extent waxing. The Chinese seismological program, which had long invested heavily in prediction research, achieved a seemingly stunning success with its prediction of the 1975 earthquake that struck near the city of Haicheng. This prediction, which led to widespread evacuations from dwellings that collapsed when the earthquake struck, saved thousands of lives. (Haicheng was at the time home to about a million people.) Yet, as Richter observed, "They were fortunate, in that the preliminary signs were far more definite than usual." By this he meant that in the months prior to the magnitude 7.4 mainshock on February 4, 1975, the region was rocked by an increasingly energetic sequence of smaller foreshocks. Those who are inclined to believe that animals can sense impending earthquakes often point to this event: unusual behavior was reported on the part of many animals, including fish and reptiles. However, the skeptic would point to the energetic sequence of foreshocks, which by itself would have surely been sufficient to sound an alarm to man and beast alike.

The idea that animals can sense impending earthquakes remains a persistent myth among the public. After any large earthquake, some people are always convinced that their dog, cat, or rose-crested cockatoo knew the temblor was coming. The problem is that on any given day, a certain percentage of cats act bizarrely for no apparent reason, because that's what cats *do.* When an earthquake strikes, anecdotal accounts of prescient animals invariably arise. Animals can also sense an initial P wave that escapes the attentions of human observers, and therefore sometimes react a few seconds before the stronger S wave arrives.

One of the reasons that the myth persists is that it is not an altogether absurd to believe that animals might sense an impending earthquake. Some animals can detect magnetic fields, for example, and it is possible that the preparatory process leading up to a large earthquake might produce electromagnetic signals. One cannot, however, draw any conclusions about animal behavior and earthquakes from the anecdotal reports that invariably surface after an earthquake. To demonstrate a link, one would have to make a far more careful, systematic attack on the problem. In fact, in the mid-1960s, a couple of clever young scientists at the Seismo Lab set out to do just this. One student tended a tank full of electric eels and monitored both their signals and earthquake activity. Another rigged up a system whereby the activity of a box of cockroaches could be measured according to how often their feet touched a sensor on the floor, and compared this data to earthquake activity. Neither these experiments, nor any of the others that scientists have done over the years, revealed any correlation between the antics of critters and unrest in the earth. It is, on the other hand, clear that some animals are very sensitive to vibrations, and thus can react to foreshocks that might not be noticed by nearby humans.

The Chinese earthquake prediction record also began to look less compelling. Although the warning signs prior to Haicheng were, in Richter's words, especially "definite," they were more typically indefinite a year later when a magnitude 7.8 earthquake occurred in northern China. Heralded by neither a dramatic foreshock sequence nor a prediction, this temblor struck the large city of Tangshan, a stone's throw from Beijing. The official death toll is listed as 250,000; many believe it was as much as three times higher.

While these events played out on the far side of the planet, seismologists in the United States and elsewhere felt a wave of optimism that observable, physical changes in the earth might signal impending large earthquakes. A number of studies by both American and Russian seismologists appeared to reveal that, prior to a large earthquake, the properties of rocks surrounding the impending earthquake changed in such a way that changes the ratio between the speed of P waves and that of S waves. Laboratory investigations of wave speeds in rock samples provided a physical basis for this observation: as rocks are put under increasing pressure, as they might be prior to a large earthquake, small cracks form within the rock, causing its volume to increase slightly. Such a process will have an effect on the speed of seismic waves that happen to be passing through; the speed of P and S waves will be affected differently and so the ratio between the two will change.

Prediction methods based on changes in seismic wave speeds began to gain traction in the 1970s; as it happened, coinciding with another development that would go on to even greater infamy in the annals of earthquake prediction research, the Case of the Palmdale Bulge. And so we return our attentions to this part of the story, for which just a bit more background is required. At the time of the 1964 earthquake, earthquake monitoring within the U.S. government was under the auspices of the Coast and Geodetic Survey. "And then suddenly the Geological Survey got interested," Richter said in his 1979 interview, in launching projects that eventually succeeded "in a very complicated way in practically elbowing the Coast and Geodetic Survey out of the picture." Richter's views of this transition shined clearly through his comments regarding the older USGS geologists who suddenly "felt qualified to make very pontifical assertions about highly critical points in seismology, and in some cases that caused a great deal of unnecessary expenditure of public funds." He also decried a USGS policy of publishing final results very late but also allowing preliminary "Open-File Reports" to be published without adequate critical peer review.

Richter's scorn for USGS geologists was largely reserved for geologist Bob Castle and colleagues. In the early 1970s Castle analyzed old surveying data and concluded that a large tract of the desert floor north of Los Angeles had risen by as much as ten inches in the years prior to 1975. This would be the infamous Palmdale Bulge, the answer to the question that George Alexander posed in a 1976 *Popular Science* article: "What is 10 inches high in some places, covers more than 4500 square miles, and worries the hell out of laymen and professionals alike?" The reason for the worry was not difficult to understand: the ominous, substantial warping of the earth's crust had taken place immediately adjacent to a segment of the San Andreas Fault that last produced a great earthquake in 1857. This huge new bulge might just be the harbinger than scientists had been watching for for decades, the sign that the next Big One was imminent.

Castle's intriguing result launched a full-scale assault to further understand the observation. In 1976 President Ford authorized two million dollars in earthquake prediction research focused on the Palmdale Bulge. A year later, analysis revealed that the bulge had grown to 32,400 square miles, with a maximum height of twelve inches. Yet the new results were perplexing as well as intriguing: further analysis and field surveys revealed that some parts of the so-called bulge had actually dropped.

Then the bulge began to develop cracks. Geophysicists Rob Reilinger, Larry Brown, and Bill Strange pointed to possible errors that might have colored Castle's results. Two seismologists at UCLA, Professor David Jackson and postdoc Wook Bae Lee, pursued these ideas in more depth. They looked at the data and found evidence that systematic errors had contaminated the leveling data. Such errors were by no means uncommon in early leveling surveys. Nowadays scientists can measure relative positions on the earth's surface with great precision using the global positioning system, but measuring relative positions to within a few centimeters was difficult with traditional surveying methods. Jackson and Lee discussed two possible sources of error, one caused by a failure to account for the fact that the length of the metal rods used in the survey depended (slightly but systematically) on the temperature. The second source stemmed from the fact that the survey lines were made from points at markedly different elevations, as low as sea level and as high as three thousand feet. Survey lines thus traversed air that changed temperature systematically along the way, giving rise to a bending of light energy known as refraction. Scientists were well aware of refraction errors by the mid-1970s but were caught off guard by the severity of the error in this particular case. Most leveling lines were done on relatively level ground and in mild climates; few involved the large elevation and temperature changes that one encounters between the southernmost Mojave desert floor and the San Gabriel Mountains.

The 1979 annual meeting of the American Geophysical Union included a session that would later be remembered as the Battle of the Bulge. Lee and Jackson presented detailed "rod calibration" errors and discussed possible additional biases from refraction. U.S. Geological Survey scientists fired back with their own interpretations and rebuttals. Geophysicist Nancy King, who had no stake in the debate at the time, recalls wondering if the heated discussions would lead to actual fisticuffs. They didn't, but that the thought crossed her mind reflects a level of animosity far outside the norm for scientific discourse.

The USGS scientists—those who spoke up—presented a united front at the 1979 meeting, but geophysicist Ross Stein would later break ranks. Although he published a paper in 1981 concluding that measurement error could not account for the observed uplift, he later investigated the rod calibration errors in more detail and found them to be substantial; by the time he was done the bulge had been reduced to half its former size. By then, Rob Reilinger had discovered that another part of the apparent bulge was in fact caused by the fact that its elevation had been calculated relative to a supposedly fixed reference point in Saugus, but Saugus had itself dropped due to groundwater extraction. Correcting for this all but erased the bulge entirely. Stein ran a field experiment designed specifically to understand the effects of these errors.

Throughout these years tensions continued to run high. Ross Stein twice found paper bags containing dog excrement in his USGS mailbox. For his part, Jackson observes that "some people told me I was full of crap, but no one actually sent me any."

By 1981, Dick Kerr wrote an article in *Science* titled, "Palmdale Bulge doubts now taken seriously." By the mid-1980s the Palmdale Bulge was looking more like the Palmdale soufflé—flattened almost entirely by careful analysis of the data. When all of the errors were corrected, the data revealed only a very small uplift of a limited area, probably the lingering effects of the 1952 Kern County and/or 1971 San Fernando earthquakes. According to Stein, a younger colleague, Gerald Bawden, eventually came along and "did some good science with what was left [of the bulge]—redemption for us all." The memories of shouting matches and dog excrement fading, Stein can also look back at his experiences with a measure of equanimity—with a measure of appreciation, even, for the lessons it taught him. "We are almost always chasing discoveries at the edge of our measurement abilities," he observes, "so the fragility of our inferences must always be borne in mind."

The Palmdale Bulge did not disappear from public and scientific concern overnight, but by the mid-1980s even its most fervent supporters had thrown in the towel. Yet the die had been cast. The mid-1970s had been heady days for advocates of earthquake prediction programs: in 1976 seismologist James Whitcomb, at the time a research fellow at Caltech, went public with a specific prediction that a magnitude 5.5–6.5 earthquake would strike within or north of the greater Los Angeles region before April 1977. Recognizing the science to be immature, Whitcomb himself was careful to refer to a "hypothesis test" rather than a "prediction." The public, however, did not appreciate the distinction. His hypothesis was based on theories involving changes in earthquake wave speeds, not on the Palmdale Bulge itself, but, ominously, the region over which the changes had been (apparently) observed overlapped with the (apparent) bulge. Whitcomb and his hypothesis made an understandable splash in the public arena: in May 1976 *People* magazine ran a three-page article describing the prediction in detail. In George Alexander's words, "Earthquake prediction, long treated as the seismological family's weird uncle, has in the last few years become everyone's favorite nephew."

The nephew never quite lived up to expectations. Whitcomb's earthquake never happened, and the Palmdale Bulge deflated into quiet obscurity. The business of earthquake prediction fell seriously out of favor within the seismological community; the family began to view the uncle as not only weird but highly suspect as well. Still, the legacy of this tumultuous era was, in the end, by no means entirely negative. In 1977, the United States launched the so-called National Earthquake Hazards Reduction Program, known to its friends as NEHRP ("knee-herp"), which represented a significant increase in the government's commitment to research aimed at reducing earthquake hazards. Although by no means sold exclusively based on earthquake prediction, let alone based solely on the Palmdale Bulge, prediction was a key element of this program. Several decades later, NEHRP continues to fund the lion's share of such research by scientists with the U.S. Geological Survey and other government agencies, as well as in academia. The program can point to a number of real successes, accomplishments as far-flung as a better understanding of the 1811–12 New Madrid sequence of earthquakes in the North American midcontinent and the "real-time" systems that now make sophisticated results available not just to scientists but also to the public and emergency managers, within a few minutes of a significant earthquake.

The NEHRP program did not, however, pave the way to earthquake prediction, as many seismologists had hoped in the optimistic days of the 1970s. Jackson, Strange, Stein, and others had deflated the once-ominous Palmdale Bulge. The rock velocity results were eventually discredited as well, not as fraudulent in any way but rather as overly naive due to a fundamental limitation that plagues almost all seismology studies: seismologists generally cannot do the kind of repeatable experiments that allow other scientists to verify their results. For example, when scientists estimate the speed at which seismic waves travel in any part of the crust, they usually rely on the waves generated by earthquakes. Artificial explosions can be used as well, but they are both expensive and, as a rule, unable to probe very deep into the crust. If, then, one wants to investigate how the speed of waves changes in time, a seismologist must analyze data from an ever-changing suite of earthquakes—and herein lies the rub. The changes in wave speeds inferred by the 1970s studies were all very small, and later studies showed that they could all be accounted for by the fact that different sets of earthquakes probed slightly different parts of the crust.

One solution to this conundrum was to turn to explosions—everything from quarry blasts to nuclear tests—as a more repeatable source of seismic waves. These studies revealed no detectable change in wave speeds. Later seismologists developed clever methods based on the observation of so-called repeatable earthquakes. It turns out that, along some patches of some faults, small earthquakes recur like clockwork, producing nearly identical seismograms time after time after time. Conceptually, repeating earthquakes are thought to be generated by small, strong patches of a fault that are surrounded by larger patches that move slowly and gradually via a process scientists know as fault creep. The small stuck patches remain stuck for a certain amount of time before their inherent strength is overcome and they come unstuck. Repeating earthquake studies corroborated the results from the earlier explosion studies, and negated the apparently promising results from the early earthquake studies.

And earthquake prediction was back to square one. Again.

Although Richter's career ended too soon to witness the discovery of repeating earthquakes, one is inclined to think that, exhaustively careful observationalist that he was, he would have appreciated them. Indeed, one has to wonder if he wouldn't have pioneered such studies himself if repeating earthquakes occurred commonly in Southern California, but they don't. Repeating earthquakes are observed on faults that are able to creep steadily rather than move only in earthquakes, for example parts of the Hayward Fault in the East Bay Area and a central stretch of the San Andreas Fault near Hollister. Southern California turns out to be conspicuously lacking in creeping faults, and therefore lacking in repeating earthquakes.

Although Richter caught only the beginnings of the chapter of the earthquake prediction story that involved changes in wave speeds, he was around for any number of earlier iterations by amateurs. He described recent work on the part of "hundreds of would-be predictors." "Every year or so," he wrote further, "one of them gets unwarranted attention from the news media. The pattern is always the same: claim to have predicted some earthquake months or years past; claims of recent successes, without listing failures; new predictions for the future." And the pattern, in Richter's estimation, continued, with, "claims that these predictions have been fulfilled by seizing on almost any earthquake anywhere near the suggested time, and ignoring larger events which do not fit."

Interestingly, Richter remained open-minded about some of the ideas that would-be amateur predictors have long touted as having potential for earthquake prediction. In a 1958 letter, for example, he mentioned the possibility that the movement of large air masses might influence earthquake activity, although he also noted that no such correlation had been proven. "There is," he wrote, "a distinct tendency for the number of small earthquakes in Southern California to increase slightly at the end of the summer, about the time of the first good rains. I have generally attributed this rather to the shift of air masses southward than to actual rain." While it remains unproven, the possibility that large atmospheric pressure changes might play a role in triggering earthquakes is not outside the realm of possibility. The earth's atmosphere creates a pressure of 14.7 pounds per square inch on average, defined to be 1 *bar.* The most extreme weather conditions can cause pressure changes on the order of 0.1 *bar*; fairly routine pressure fronts can cause pressure to increase or decrease by a few percent of a *bar.* According to recent seismology research, even these smaller changes might, as Richter speculated, be sufficient to influence earthquakes. The notion that earthquakes might be correlated to certain weather patterns is thus not implausible. Where Richter differed from many amateur predictors was in his understanding that any such correlation—if in fact it is real (this remains unproven)—was very likely to be weak at best, and of little to no practical use in prediction.

Richter's words on earthquake prediction resonate with uncanny truth and prescience to the modern seismologist who can bring to mind recent examples of this pattern. In one recent iteration, a team of individuals, including one with a Ph.D. in earthquake science, developed a Web-based business that promised to predict earthquakes—for a nominal monthly fee. They touted an impressive rate of success predicting past earthquakes (without listing failures). Their new predictions, examined closely, revealed the aforementioned liberties with the "useful narrow windows" part of the equation. They would, for example, predict that a magnitude 2.5 or greater earthquake would occur within a given week near the California-Mexico border. Such an event was very likely to occur because that region generates a very high rate of small earthquakes. Worse still, these would-be predictors gave themselves a "hit" if an earthquake of the "right magnitude" struck anywhere within hundreds of kilometers of the specified location—and "right magnitude" spanned a very large range. They could thus predict a magnitude 2.5 or greater earthquake near the California-Mexico border—

an event that was very likely to occur—and consider it a success if a magnitude 6 earthquake struck near Palm Springs. In the technical parlance of science, this is what we call shooting fish in a barrel.

Even established scientists sometimes take liberties, for example claiming to have made successful predictions based on research that only quantifies regions where earthquakes were likely to occur based on short- or long-term patterns of earthquake occurrence—without giving any specific windows in time or magnitude. In weather-forecasting terms, this is not shooting fish in a barrel but rather akin to claiming to have predicted a particular tornado in central Kansas based on the assessment that tornadoes are likely in the Midwest. Charles Richter would have been apoplectic.

In his unpublished 1976 notes, Richter got it all off his chest, so to speak, writing with such candor as to feel compelled to add a disclaimer at the beginning, "This is a personal statement, in no way official, and not to be attributed to Caltech or any of its staff."

One cannot quite tell whether he was referring to amateur or scientific prediction efforts in the second to last paragraph of his notes. One suspects the former, but cannot be sure: "A few such persons are mentally unbalanced; but most of them are sane—at least in the clinical or legal sense, since they are not dangerous, and are not running around with bombs or guns. What ails them is exaggerated ego plus imperfect or ineffective education, so that they have not absorbed one of the fundamental rules of science—self-criticism. Their wish for attention distorts their perception of facts, and sometimes leads them on into actual lying."

One begins to understand why he felt compelled to add his disclaimer.

Richter's most damning words were most likely meant for amateur earthquake predictors rather than his professional colleagues. His notes did close with a somewhat more conciliatory observation: "Occasionally a professional man who has a good reputation in other fields is responsible for erroneous statements about earthquake occurrence and earthquake prediction. Even good geologists have been known to fall into such errors."

The history of earthquake prediction research within the mainstream scientific community in recent decades reveals a pattern: promising initial results hint at signals in the earth that preceded one or more large earthquakes in the past, but these signals do not stand up to closer scrutiny—or they fail to occur before the next large earthquake in a given region. It seems almost self-evident that dramatic changes would take place on or near a fault as a large earthquake on that fault becomes imminent. These changes, one reasons further, should be detectable as either changes in the properties of underground rocks or in the patterns of small earthquakes.

This latter thought motivated Harry Wood during the very earliest days of the Seismo Lab: as he endeavored to build an earthquake-monitoring program from scratch, he was driven in part by the belief that studying small earthquake locations would provide clues to where future large earthquakes would occur. Recall that in his "top secret" letter to Bob Sharp, in which Richter provided a detailed overview of lab operations, Richter had written, "There was original hope that little shocks would cluster along the active faults and perhaps increase in frequency as a sign of the wrath to come. Too bad, but they don't. Roughly, little shocks on little faults, all over the map, any time; big shocks on big faults." This observation was presented and supported more formally by a seminal 1965 paper on which Richter's longtime collaborator Clarence Allen was lead author.

Seismologists would generally agree with Allen's and Richter's assessment: the steady spattering of small shocks appears not to provide any clues to the timing of future large earthquakes. Only two possible exceptions have been proposed. First, some large earthquakes are preceded by smaller foreshocks, typically within a few days of the mainshock. It has been, however, impossible to distinguish foreshocks from garden variety small earthquakes, which means that we don't know that an earthquake is a foreshock until after the mainshock occurs. The second exception, and this remains open to debate, is the possibility discussed in earlier chapters that moderate shocks occur more commonly in a region in the years or decades prior to a major earthquake. This observation leads some seismologists to think that intermediate-term prediction might some day be possible: the identification of regions where a big earthquake is likely to occur in the next few years or decades.

The field of seismology has seen many important developments in the decades following Richter's death. Among the most notable is a better understanding of the interaction between earthquakes, not just mainshocks and aftershocks but the way that one large earthquake can influence the timing of another. We also, as previously noted, have a better understanding, or so some believe, of the so-called earthquake cycle. Short-term earthquake prediction, however, has proven a tough nut to crack.

Efforts continue. In early 2004 a team of scientists from the University of California at Los Angeles led by Vladimir Keilis-Borok garnered substantial media attention with their prediction that a large temblor would strike the Southern California desert within the first nine months of that year. Their prediction was based on a twofold pattern that they concluded had preceded other large earthquakes in California and elsewhere. The first part of the pattern was familiar to seismologists: an increase in moderate shocks in the years prior to a large mainshock. But Keilis-Borok's team identified a new precursory pattern as well: the occurrence of a series of earthquakes that were closely spaced in time and stretched out spatially to form a so-called chain. An earthquake chain might comprise a half-dozen magnitude 3–4 shocks stretching out over a few hundred kilometers. It is not clear what would cause these chains to form, but Keilis-Borok offered several suggestions, including a process deep in the earth that causes a given fault to weaken, for example because of fluid migration.

Keilis-Borok's group had circulated private predictions prior to large earthquakes in Japan in 2003 as well as the magnitude 6.4 San Simeon earthquake that struck central California in December of that same year. Seismologists couldn't help but wonder: were they onto something? For perhaps the first time, a group of serious seismologists were proposing a prediction method, and a specific prediction, that could not be dismissed as out of hand. When Keilis-Borok presented his prediction at the annual meeting of the Seismological Society of America in April 2004, many of the scientists in the audience had their doubts, but nobody in the audience budged during a talk than ran four times the usual length. In the end, 2004 came and went with considerable heartburn for those who live in the Southern California desert but without so much as a moderately damaging earthquake in the region. The prediction scored a "miss," but a single failure does not by itself discredit the method. Only by making a series of predictions over a period of many years will scientists be able to tell whether or not the method has any validity. (As of mid 2006, the track record of the method, as estimated from a dozen or so predictions, is not looking good.)

Other recent research has focused on so-called slow earthquakes and a kind of jittery earthquake chatter, both of which have been detected in a transitional region in the earth's crust, below the faults that can generate earthquakes and above deeper layers that are more plastic. The slow earthquakes have been observed in a number of subduction zones, along the interface between the sinking oceanic plate and the overriding continental plate. The chatter, or tremor, has been observed below the central San Andreas Fault near Parkfield. Once again the seismological community is abuzz with, if not excitement, at least *possibilities*: could these deeper disturbances provide the nudge that finally causes a highly stressed fault above to break in an earthquake?

Among the most vexing aspects of earthquake science is this: we don't know what we don't know. As Richter himself put it in an interview with Henry Spall in 1980, "Nothing is less predictable than the development of an active scientific field." Wisely, he was unwilling to even speculate on the likelihood of reliable earthquake prediction in the future. Eric Lindvall emphasizes that Richter was never *against* earthquake prediction per se, but rather remained enormously skeptical—based on his own professional expertise—that seismologists were close to developing a reliable prediction method.

At the present time, as in the waning years of Richter's career, seismologists have little reason to believe that a reliable short-term prediction method is on the horizon. This unfortunately will do little to dissuade the "fools and charlatans" who will always be among us. Many of Richter's observations, even the more caustic ones that most of us would never dare utter (even though we might wish to), ring all too true to modern ears. But science moves forward in strange and often serendipitous ways. Can we predict earthquakes now? No. Can we predict whether or not prediction will ever be possible? No. These answers are as true today as they were in 1978. But can we say with certainty that prediction will never be possible? The answer that Richter gave decades ago is as good as the answer that seismologists can give today: we don't know. This uncertainty leaves us precisely where Richter found himself decades ago: forced to draw on integrity to resist false hopes, however strongly they may beckon.

CHAPTER 17

Sizing Up Earthquake Hazard

The idea of earthquake prediction appeals to the imagination, and attracts disproportionate attention from the public, the news media, and some officials. The immediate objective of reducing earthquake risk to lives and property would be best served by the removal of old dangerous buildings.

—*Charles Richter, memo, 1976*

No SEISMOLOGIST could have witnessed the aftermath of the 1925 Santa Barbara and 1933 Long Beach earthquakes without realizing what seismologist Nick Ambraseys would later sum up succinctly as follows: earthquakes don't kill people, buildings kill people. As the two early Southern California temblors demonstrated—and far too many moderate temblors around the world continue to demonstrate even today—poorly designed and/or poorly built buildings kill people.

In areas where wood is scarce or prohibitively expensive, it is especially challenging to build buildings that can withstand earthquakes. Unreinforced masonry construction, which one finds ubiquitously in older and poorer cities in many countries around the world, has little ability to withstand the shaking from even relatively modest temblors. Thus, when a magnitude 6.6 earthquake struck the city of Bam, Iran, on December 26, 2003, mud-brick buildings in the ancient city crumbled, killing over twenty-six thousand residents of the city. Just four days earlier, on December 22, 2003, an earthquake of nearly comparable magnitude (6.4) struck near the central California city of Paso Robles. The epicenter of this quake was not directly beneath a large city but rather in a remote and hilly region. Without question this resulted in less damage than would have occurred had the temblor occurred in an urban area. Still, Paso Robles and neighboring cities experienced severe shaking—shaking that would have been devastating in a city like Bam but was far less destructive in a modern California town. A few old masonry buildings in downtown Paso Robles collapsed, killing two people—a tiny fraction of the death toll in Bam.

In and around Paso Robles chimneys fell, wine barrels tumbled, and kitchens and grocery store isles were left in shambles. But the vast majority of homes and other structures withstood the assault with, at worst, modest damage, because commercial and public buildings in California have for decades been built to stringent earthquake codes. Homes and smaller apartment buildings are not only built to strict codes as well, but are also almost always constructed of wood. The availability of wood as an affordable building material has been a fortunate happenstance for residents of California. Even without special design, the average well-built wood-frame house tends to ride out earthquake shaking relatively well. Wood-frame houses can be vulnerable to damage in some cases, for example if they are built with elevated cripple walls—as was the unfortunate common practice with the craftsman bungalows that are common in some older communities. Wood-frame houses can also get knocked off their foundations: a problem that is solved by, literally, bolting the main structure of a house to its foundation.

Between abundant wood for building and stringent building codes that have been in place for decades, California's "earthquake problem" has been substantially reduced. The rate of earthquakes might be high, but this represents what a scientist terms the earthquake *hazard*. What matters to life and limb, however, is earthquake *risk*: the vulnerability of structures and lifelines to the hazard. It has been through tireless efforts on the part of seismologists, engineers, public officials, and others that this risk has been largely ameliorated—especially compared to other parts of the world that are less well prepared.

Largely ameliorated, however, does not mean entirely ameliorated—as Richter recognized as a young scientist in the 1930s and continued to talk about throughout his career. Recall colleague Robert Sharp's observation, "He was very compassionate about the people living in those buildings, and he had a strong sense of community with people. He just had trouble expressing those feelings."

No seismologist whose research involves hazards can avoid the sense of profound frustration that comes with the realization that proper design and construction can solve the "earthquake problem," perhaps not to the extent of avoiding all loss of life, but certainly to a very large extent. And yet the pattern plays out over and over: virtually every large earthquake in California claims more of the state's old buildings as well as some relatively new ones of a design that has proven to be more vulnerable to damage than first thought. "The general picture of conspicuous damage," Richter wrote after the 1971 earthquake, "was of familiar character. Masonry buildings along the principal street in San Fernando presented appearances easily matched from 1933 or 1940 or 1952. As was done in Bakersfield after the earthquakes of 1952, many three and four story brick structures are being cut down to one or two stories."

Although California has some of the most stringent building codes in the world, they apply to new construction only (with the exception of older public school buildings, for which the Field Act does include provisions). No state law has ever been passed to mandate the retrofitting of structures that are now recognized to be vulnerable to earthquake damage. (Some cities, however, have mandated rudimentary retrofitting of certain older commercial structures.) Even without legislation, homeowners are sometimes motivated to take steps to protect what is often their largest investment, for example by adding plywood bracing to cripple walls or bolting an older house to its foundation.

Commercial buildings are often a different story. Retrofitting a commercial building can be an expensive, more expensive than designing it properly in the first place. Consider existing older brick commercial structures in a community such as Pasadena. Space in these buildings might be leased to small businesses that do not pull in huge revenues: antique shops, coffee shops, yoga studios. A building owner would have to shoulder the cost of retrofitting without being able to raise rents afterward, not, as a rule, an economically viable proposition.

Worse yet, under the current rules of the game, retrofitting commercial buildings is often not an economically advantageous proposition. If an owner of commercial property in California realizes that a building is vulnerable to earthquake damage, he or she has two choices: pay for the expensive retrofit, or do nothing. The former option requires a substantial outlay of funds for no immediate benefit. If the latter option is chosen, then the earth itself has two choices: it can remain quiet and dish up no large earthquake nearby, or it can generate a large enough and close enough temblor to cause serious damage. The former option is scarcely unlikely, even in earthquake country, during the typical length of ownership of a commercial building. Thus the "do nothing and hope for the best" approach works to a property owner's advantage in a great many cases. Moreover, if a property owner chooses to do nothing and his or her hopes for the best are dashed, government assistance will be available.

And herein lies the stark reality at the heart of the problem: as a society we have constructed a system in which it is not necessarily in a commercial property owner's interest to make their building earthquake-safe. An owner incurs no liability under current law by failing to retrofit a structure that is known to be highly vulnerable. Moreover, the owner trusts that disaster relief will be forthcoming if a damaging earthquake does strike. Over two hundred years ago, Adam Smith explained how "rational self-interest" in a free-market society will lead to an overall thriving economy. Yet Smith also understood the inability of a free-market economy to work to the direct benefit of the public good. In earthquake country in the present day and age, "rational self-interest" alone leads, ironically and tragically, to the continued existence of buildings that could very well kill people some day.

For new commercial construction, the implementation of earthquake-resistant design is sometimes said to add an extra expense that is "less than the cost of the carpets," perhaps a few percent of the overall construction expense. Retrofitting an existing building is far more expensive—an argument for building things right in the first place. And while the costs of retrofitting can be prohibitive, so too are the staggering costs of putting a city back together after a major disaster—to say nothing of human lives that can never be put back together.

Thus, where earthquake safety is concerned, we have a fundamental disconnect between enlightened self-interest of individual property owners and the collective self-interest of society. We have developed a haphazard patchwork quilt of federal disaster relief policies and state and local building codes and practices—vulnerable buildings remain vulnerable because it is financially advantageous for property owners to take their chances. Even if they roll the dice and they lose, they don't really lose—and yet we all lose.

Nowhere in Richter's writings does he reflect on these issues. Richter saw the equation from a purely scientific—not to mention a wholly unpragmatic—point of view: old buildings are a menace, old buildings should be fixed or torn down. (He further railed against skyscrapers, which he saw as an unnecessary risk. Many in the engineering community regarded his views as naive. According to Eric Lindvall, later in Richter's life he did concede that perhaps tall buildings could be safe, if properly engineered.) It remains today, as it was several decades ago, a continuing source of vexation to seismologists that, first, the problem has a solution and, second, that the solution is desirable from a collective, societal viewpoint—and yet the solution is not implemented.

Richter's ardent, lifelong criticism of earthquake prediction by no means reflected a lack of interest in the idea of predicting earthquakes, but rather an understanding of how much shoddy science had been done in scientists' quest for prediction (and how much funding had been diverted as a result into pointless endeavor), and the fact that it distracted people from effective hazard mitigation.

As earlier chapters discussed, Richter remained largely "California-centric" throughout his personal and his professional life. This chapter follows his lead, addressing earthquake hazard issues particular to the Golden State. Richter was, of course, well aware of earthquake hazard in other parts of the United States and the world. A long chapter of *Elementary Seismology* is devoted to California earthquakes; another long chapter discusses earthquakes in New Zealand, one of the closest analogs, with respect to earthquakes, to California anywhere in the world. (Turkey is another.) Richter's visits to New Zealand had given him a chance to spend several weeks touring the country's spectacular faults.

Richter's career ended a few years too soon to witness one of the more important developments in hazards-related seismology research in the United States. In 1984, two of his former colleagues at Caltech, Tom Heaton and Hiroo Kanamori, published a paper titled "Seismic Potential Associated with Subduction in the Northwestern United States." This paper considered the Cascadia subduction zone, where the oceanic plate sinks beneath the North American continent along the coasts of Oregon and Washington. Cascadia is similar in a general sense to the subduction zone along the eastern edge of the Indian Ocean that generated the devastating earthquake and tsunami of December 26, 2004. Prior to the mid-1980s, scientists debated whether the Cascadia zone produced great earthquakes, or if it was among the handful of subduction zones around the world where oceanic crust sinks gradually, without occasional great earthquakes.

Heaton and Kanamori argued that the Cascadia zone had much more in common with zones where great earthquakes do occur than with their more benign brethren. A few short years later, geologist Brian Atwater found the first geologic evidence that a large earthquake and tsunami had struck the coast in the not-too-distant past. Within two decades, scientists had determined not only the precise date of the last great Cascadia quake (January 26, 1700) and time (nine o'clock at night, local time), but had also found evidence that previous great earthquakes have occurred about every five hundred years on average. (For those inclined to take comfort in the fact that it has been "only" three hundred–odd years since the last one, don't get too comfortable: the geologic record reveals that the earthquakes do not recur like clockwork, but can strike as much as eight hundred years and as little as two hundred years apart.)

Richter's concerns for old and vulnerable buildings in California reflected his career-long focus on earthquakes and earthquake hazard close to home. Had he been alive to witness the renaissance in thinking about earthquake hazards in the Pacific Northwest, one suspects his concern for vulnerable old buildings might have expanded. California's vulnerable old buildings do pose a problem, but it pales in comparison with other parts of the country (not to mention the world) where large or even great earthquakes can happen and yet building codes have lagged far behind California's in incorporating earthquake provisions.

In Richter's day as now, people sometimes marveled how anyone could live in California, a state that clearly could not sit still. There is an answer to this question; a sensible answer, even. No place on earth is guaranteed to sit still and stay still forever. If one is going to live on a dynamic planet, best to pick a place that has had the sense to prepare itself.

For all of his crusades for hazard mitigation, Richter did have an appreciation for the issues of *relative* risk. Human can be quite irrational about such matters, for example fearing an airplane crash far more than the much bigger risk of an automobile crash. In an average year, over forty thousand Americans lose their lives as a result of automobile accidents. We accept this risk in part because we think it is within our control: if I drive carefully, it won't happen to me. Yet the numbers are staggering if one does the math: during the span of an eighty-year lifetime, 3.2 million people will die in car accidents, a number equal to about 1 percent of the U.S. population. Put another way, the average American stands a one in one hundred chance of meeting his or her death in an car accident. In contrast, the sum and total of all earthquake fatalities in the United States, ever, barely adds up to a few thousand.

Richter understood the concept of relative risk. "I don't know why people in California or anywhere else worry so much about earthquakes," he once said. "They are such a small hazard compared to things like traffic." His closing observation to writer Digby Diehl was, "Given the choice, I'd worry about the freeway a lot more than an earthquake any day." Although those who heard these sorts of observations might have assumed they were meant in jest, Richter was not joking. In other countries the story is different: large earthquakes have claimed staggering tolls in countries such as China, Iran, India, and Turkey, where building codes and construction are woefully inadequate for the earthquake hazard. In some parts of the United

States, even a moderately large earthquake could claim a heavy toll. Imagine, if you will, a magnitude 6.5 earthquake striking downtown Boston. Such a temblor would be unlikely, but prior to late 2004, most seismologists would have said that a magnitude 9.0 quake along the eastern edge of the Indian Ocean was unlikely too. As a society we prepare ourselves for the probable, but sometimes improbable things happen.

In California, the building codes and standards have been strict for a very long time. Even as he led the charge to improve the situation further, Richter knew this. One hopes he also felt a measure of personal gratification. It was, after all, largely by virtue of the pioneering efforts of the first generation of Southern California seismologists—of which Richter himself was a charter member—that Earthquake Country is today as safe as it is.

CHAPTER 18

Hazard in a Nuclear Age

As a student I was much interested in the attractive possibilities of physical and biological science that seemed then to be in the near future—such as the exploration of space; the unraveling of atomic structure and the rationalization of chemistry; the release of nuclear energy; the development of what we now call molecular biology.
—*Charles Richter, 1970*

DURING the latter half of the twentieth century seismologists found their expertise tapped for two key public policy issues of the nuclear age: test ban treaty verification and the safety of nuclear power plants. The former issue proved to be a boon to global earthquake studies. As policymakers came to understand that the best way to monitor nuclear tests—ours as well as those by other countries—was to record the waves they generate within the earth, global earthquake-monitoring networks became a priority for not only scientists but also for politicians. The World-Wide Standardized Seismographic Network (WWSSN) was born in the early 1960s, funded not by the National Science Foundation but rather by the Advanced Research Projects Agency, the research arm of the Department of Defense.

Nuclear power plant safety, on the other hand, involved local rather than global studies. Prior to the development of nuclear reactors, structures such as schools and hospitals were the critical facilities to which engineers and emergency managers devoted special attention. When scientists split the atom, they upped the ante. The Nuclear Regulatory Commission (NRC) had to consider the possibility that earthquakes could cause damage to reactors, leading to a catastrophic release of radioactive elements. This eventually led to pitched battles over California's two currently operating commercial power plants: Diablo Canyon near San Luis Obispo on the central California coast and San Onofre, on the coast between Los Angeles and San Diego. Both plants were built over opposition from scientists and citizens who questioned the wisdom of building nuclear facilities in the heart of an active plate boundary. That earthquake hazard along California's immediate coastal corridor is lower than the hazard along the state's major fault lines is beyond dispute. The key question was, and remains, how much lower? Without question, active faults run along both California's coastal regions and its offshore borderlands. Without question, all of California is earthquake country.

Diablo Canyon and San Onofre were, however, designed and built with an awareness of earthquake hazard. The magnitude 6.4 San Simeon earthquake of December 22, 2003, struck just fifty kilometers from the Diablo Canyon plant. The site experienced fairly severe shaking, but the structure had been designed to survive much stronger shaking from a magnitude 7.2 earthquake on the nearby Hosgri Fault, and rode out the 2003 temblor unscathed.

The story of California's nuclear power plants—their degree of earthquake safety in particular—in fact traces its roots back to the 1970s, to battles fought in very different geologic settings. When power companies sought licenses for nuclear plants in less seismically active parts of the United States, earthquake hazard merited less concern. Through the 1970s companies such as Con Edison in New York hired consultants who considered local earthquake hazard in places where it did not generally rank high on anyone's radar screen. According to NRC policy, the onus was on companies to evaluate earthquake hazard, including determining whether or not a proposed plant was in proximity to what they defined as a "capable" fault. The full definition of a capable fault was complicated. Essentially the criteria were meant to identify those faults on which a large, shallow earthquake could plausibly occur in the near future. If no capable faults existed close to a site, the region was to be zoned into geologic provinces within which earthquake hazard could be considered uniform. A plant then had to be designed to withstand the largest historic earthquake that had occurred within that same zone.

In the 1970s Con Edison set its sights on Indian Point, a location along the Hudson River in Westchester County, New York. Although relatively sparsely populated at the time, the site's proximity to New York City—about twenty-five miles—raised concern among the public, environmental groups, and public officials. In licensing hearings before the NRC, the state chose to contest two specific earthquake-related issues: the way that Con

Fig. 18.1. Indian Point nuclear power plant. (Photo courtesy of Lamont-Doherty Earth Observatory.)

Ed's consultants had drawn their earthquake zone maps and the formula they had used in their calculations to estimate the level of shaking that would result from a recurrence of moderate earthquakes that had occurred in the area in historic times. A local environmental group contested a third issue: the consultants' assessment that the nearby Ramapo Fault was not a capable fault under NRC guidelines.

These hearings caught the attentions of seismologists at the nearby Lamont-Doherty Geological Observatory of Columbia University, across the river from Westchester County. Lynn Sykes in particular—a coauthor on a landmark 1975 paper that had first integrated seismology results into the new plate tectonics paradigm—took an interest in the proceedings. Although he brought no agenda with him to the hearings, he soon realized how steeply the deck was stacked. Con Edison brought a five-million-dollar legal war chest to the hearings; New York State had ponied up about one million dollars for their legal effort. The local environmental group, meanwhile, had a budget of about four thousand dollars. Sykes felt compelled to get involved to make sure that all of the earthquake hazard issues were discussed seriously.

Along with engineering seismologist Mihailo Trifunac, who was then at Caltech, Sykes brought substantial expertise to bear on matters that had been considered initially by Con Ed consultants. Trifunac focused his attention on the consultant's prediction of ground motions, a somewhat arcane but key issue that hinged on how one interpreted intensity data from historic earthquakes for which no seismograms were available. The consultants had cooked up one so-called intensity relationship; by the time Trifunac had shredded it thoroughly, a lawyer for New York State was heard to say that he never wanted to see that relationship ever again. Observing Con Edison's defense of seemingly indefensible science, Sykes came away with a distrust of the NRC and a determination to force a serious consideration of the Ramapo Fault.

Today, the identification of active faults in eastern North America still remains one of the most vexing challenges in earthquake hazard assessment. Seismologists who investigate earthquakes and faults in this part of the world have a saying, "Faults without earthquakes and earthquakes without faults." The identified faults in this part of the world don't seem to produce earthquakes, while sometimes large earthquakes pop up where no fault had previously been identified. One resolution of this apparent paradox is the possibility that regions like eastern North American contain a great many faults, none of which are very active but all of which are potentially active. Any given fault might stay quiet for tens of thousands of years, or more. In California it might be fair to say that an active fault is one that has been active within the last ten thousand or thirty-five thousand years; in other regions this definition might not work as well. However, given the NRC's defined protocols, the issue for the hearing was this: should the Ramapo Fault be considered capable according to their guidelines?

The Ramapo Fault is part of a system of northeast-trending faults that run from southeastern New York into eastern Pennsylvania, and possibly beyond. In California, geologists can dig into layers of sediments within creek beds to show that faults like the San Andreas have produced large earthquakes in recent geologic history. Along the Ramapo Fault one finds no such creek beds that could definitively tell whether a large earthquake has occurred within the last thirty-five thousand years. It is thus impossible to tell from geologic evidence whether or not the fault has produced significant earthquakes within the past few tens of thousands of years. According the NRC guidelines, however, the fault would still be considered "capable" if it had produced small earthquakes at a rate consistent with expectations for an active fault. This criterion relies on the familiar Gutenberg-Richter distribution of earthquakes: if a certain fault had produced, for example, ten magnitude 5 earthquakes in fifty years, one could expect that a magnitude 6 was likely to occur about once every five hundred years.

In his own testimony, Sykes pointed to a spattering of small earthquakes recorded by seismometers in the decades prior to 1976. Although not in perfect alignment with the Ramapo Fault, these small shocks generally followed the overall trend of the fault. He moreover discussed larger shocks that struck the region earlier in historic times, for example moderate (magnitude 5) shocks that rocked the New York City metropolitan area in 1737 and 1884. Sykes spent a week on the witness stand, his time not compensated, arguing against intense grilling from the Con Edison legal team.

When Sykes's testimony ended, the Con Edison lawyers brought in an expert of their own: Charles Richter, who gave his testimony on July 20 and 21, 1976. Richter testified that the Ramapo Fault could not be considered a capable fault—one that could produce an earthquake large enough to cause a surface break, upwards of magnitude 6. After pointing to the acknowledged absence of direct geologic evidence, most of his testimony and the subsequent cross-examination focused on how one should interpret the seismological evidence. Richter argued that nobody could know exactly where the historic earthquakes had struck, and that, to his knowledge, available information about the most recent moderate shock (in 1884) pointed to a location away from the Ramapo Fault. He further described results from more recent smaller earthquakes as inconclusive at best: that the locations appeared to cluster near the fault did not indicate that the fault itself was active.

Richter further focused on a more general point, that the real issue was whether or not available seismological results supported the conclusion that the fault was highly active according to the NRC's guidelines. According to the guidelines, for a fault to be considered capable based on seismological rather than geological evidence, recorded earthquake activity had to be at a certain level, a level that Richter described as "macro-seismic" (a term not in common usage today). He took exception to Sykes's earlier definition of "macro-seismic" as "an event of intensity III or more." To this notion Richter responded, "That is not only a doubtful statement; it is even ungrammatical because it correlates an abstract term with a concrete one. It is in the same category as the common joke that happiness is a warm puppy." This observation drew laughter, as did his response to a question about a large earthquake near Charleston, South Carolina, in 1886: "Counsel," he replied, "I think I see where you're driving at, and if you would shift your ground to consider the New Madrid earthquake I could probably give you an answer which would satisfy you better."

Richter earned the board's appreciation with his observation that a certain line of questioning, one focused on whether or not a certain earthquake could be considered a "principal earthquake," was "essentially irrelevant" because it did not address the central issue of whether the fault should be considered macro-seismic. To this, the chairman replied, "Dr. Richter, perhaps you should be sitting up here instead of down there because that's just about what we had concluded."

Pressed on the issue of what parts of the contiguous United States he would consider macro-seismic under the NRC guidelines, Richter responded with a short list: the region west of the Rocky Mountains, the central Mississippi Valley, the Saint Lawrence Seaway (most of which is in Canada), and possibly (but by no means certainly) the Charleston region. Sykes pressed the lawyer to press Richter further on this issue: clearly large and damaging earthquakes had struck, and therefore could strike again, in other parts of the central and eastern United States. By Sykes's recollection, the lawyer hesitated to challenge directly a scientist of Richter's stature. The lawyer did go as far as to question his degree of expertise on eastern earthquakes: "Aren't you an expert on *California* earthquakes," he asked, going on to suggest that Richter might be "prejudiced against small earthquakes we have and the patterns that might be involved with them." Richter responded that he appreciated the lawyer's meaning more than his grammar, and went on to suggest that the converse might in fact be true: "I do not mean to be facetious," he said, "but everyone tends slightly to overestimate his own earthquake in his own area. It is a little like one's own family or one's own dog."

Most seismologists today would agree that potentially damaging earthquakes, with magnitudes upwards of 6, are not restricted to the zones that Richter listed. On the specific issue under consideration—should the Ramapo Fault be considered capable under the NRC's guidelines—time has vindicated Richter. Sykes' arguments were largely predicated on a theory that had been published by his colleagues at Lamont: that the Atlantic seaboard was slowly sliding eastward away from the rest of the continent. This sliding was thought to be taking place on a number of large faults, including the Ramapo. Scientists would later learn that this interpretation had been based on data from seismometers that had not been properly wired.

When scientists analyzed more and better data, the opposite view emerged. The Atlantic Seaboard is not slip-sliding away; rather, like the rest of central and eastern North America, the coastal regions are being broadly (and rather gently) squeezed, or compressed, by plate tectonics forces. The direction of the squeezing is moreover such that the Ramapo Fault is highly unlikely to be active. Geologists might not be able to say for certain that a given fault is active, but if a fault is not oriented in the right way to respond to the direction of currently active forces, it is possible to say that the fault is not likely to move—just as a book on a table will not move sideways if one pushes it straight down.

Looking back, Sykes expresses satisfaction at what he and Trifunac accomplished against the enormously stacked deck. For the first time, power companies were put on notice that slipshod results from congenial contractors would not go uncontested. Con Ed won the immediate battle: the Indian Point reactors were designed to withstand a magnitude 5.25 earthquake, as their consultants had originally suggested was appropriate. But Sykes and Trifunac felt that they made a lasting contribution to the larger war. As Pacific Gas and Electric became interested in building nuclear power plants in California, they hired a geologist of impeccable academic and personal reputation, Lloyd Cluff, to head their earthquake risk management department. Reasonable scientists might differ in their opinion of the design criteria used for Diablo Canyon and San Onofre, but nobody could say that it had been based on cursory work.

Whether or not the Con Ed consultant's work had been perfunctory on Indian Point, no seismologist could read Richter's testimony and question his acumen. Even at age seventy-six, even as a seismologist whose entire career had been largely California-centric, he spoke the issues with singular insight and clarity, his attentions unerringly focused the matters directly before the court. When he veered onto a tangent, as he did a couple of times, it was to offer philosophical thoughts: "often we have to deal with well-meaning persons who say, well, we have to be absolutely sure, this has to be absolutely safe. And, of course, there is no absolute security in human affairs, and I need not point out that we are, all of us, constantly under the threat of very much greater risks than those against which this proceeding is directed."

The man who was almost unerringly on the side of the angels in matters related to earthquake hazard assessment and prediction-related earthquakes was, once again, correct in his answer to the question at hand: should the Ramapo Fault be considered "capable" according to NRC guidelines? The real question might be whether the *concept* of a capable fault is inappropriate in a region where large earthquakes can strike on faults that have been previously quiet for many tens of thousands of years.

Having spent his career at Lamont, Sykes's appreciation of this larger issue quite possibly did outstrip Richter's. But here once again, Richter's personality traits shone through with megawatt intensity. He answered the questions that were put before him, and he answered them with not only keen scientific insight but also with the utmost precision.

It is an interesting question whether Richter also brought an agenda to the hearings. Watching Richter's testimony, Sykes found himself wondering about his colleague's motivation: whether a scientist of that generation might not have bought into Eisenhower's "atoms for peace" initiative as a way to ameliorate lingering cultural guilt over Hiroshima. (The phrase was derived from a 1953 speech in which Eisenhower tried to portray the development of nuclear power as a benefit to mankind rather than the agent of its ultimate demise, as so many feared at the time.) Sykes is the first to acknowledge that any such speculation is only that. However, by the 1970s Richter had had the experience of living in Tokyo, during which time he developed an appreciation for the Japanese culture, people, and language. It is difficult to imagine that anyone in that situation could have failed to come away with a measure of regret over the atomic bombings. For an idealistically minded individual like Richter, it might have been a small and natural step from regret to guilt.

Without question Richter did bring some biases with him to the witness stand. This lifelong fan of science fiction genuinely embraced the vision of a clean, nuclear-powered future. As a lifelong Californian who had witnessed the very worst of the smog era in Los Angeles, he was certainly familiar with the perils of a fossil-fuel-powered present. In one of his minor tangents during the hearings he mentioned his earlier background in atomic physics and offered his view that the "development of atomic power is one of the great forward steps in the history of mankind, that it is justly comparable to the discovery and utilization of fire by primitive man, and that some risks should be taken consciously in the course of that development." In the 1970 retirement speech that he wrote but never delivered, he described the interest he had felt during his younger years for scientific developments then on the horizon, including the future "release of nuclear energy." His later regrets appear to have been focused on the advent of nuclear weaponry rather than the eventual development of nuclear power.

As the planet now faces the inevitable end of the fossil fuel era, with demands for energy increasing by the second, nuclear power beckons once again. And as the original forty-year licenses begin to expire on the first generation of U.S. power plants, their current owners are applying for twenty-year extensions. The mind boggles just a bit at the idea of nuclear power plants—the first generation of plants, no less—reaching their half-century birthdays. The mind more than boggles when one considers the stalemate over the Yucca Mountain waste repository that has left radioactive spent fuel rods without a permanent home. In the absence of a permanent waste repository, plants have been forced to store their fuel rods on site, typically in water-cooled pools. Originally meant to be only a short-term solution, plants have found their pools filling to and beyond their original capacities. With no other solution on the horizon, the NRC has allowed operators to as much as quadruple the number of rods stored in these pools.

Some highly respected scientists have challenged the safety of the storage pools, pointing to the potential for a catastrophic fire should the water accidentally be drained. Of particular concern is the growing density of the rods, and the chance that, if the water were to drain, densely packed younger rods could heat up and burn off their protective zirconium cladding, sparking a potentially runaway catastrophe. This in turn could release more radioactive material—Cesium-137 in particular—than was released during the accidents at Chernobyl in 1986 or Three Mile Island in 1979. A NRC-commissioned study estimated the potential losses: 54,000–143,000 cancer deaths, with land and other economic losses between $117 and $556 billion. Concerns for man-made disasters brought about by terrorist attack have in recent years outstripped—or at any rate greatly enhanced—previous concerns over potential natural disasters.

Alternative local storage solutions utilize either "dry casks" or a return to the older storage system known as open-racking, whereby the rods are stored underwater but in racks like giant egg crates. Were the water to drain suddenly around rods in open-racking storage, the rod cladding would remain cooled to some extent by the air. The dry cask solution involves storage of rods within cast-iron casks stored inside reinforced-concrete structures. Germany opted for this solution as its pools filled to capacity. In the eyes of some experts, the NRC has been slow to come to grips with the spent-fuel problem in the United States. A recent study commissioned by the National Academy of Sciences released their preliminary findings in April 2005, urging a return to less dense underwater storage as well as the installation of backup water-spray systems. National Academy president Bruce Alberts described the matter as a "critical national security issue."

Charles Richter lived to witness both the near-disaster at Three Mile Island and the beginnings of the Yucca Mountain project in 1978. As a proponent of nuclear power, he might well have embraced the latter as the ultimate solution for the waste issue. He could probably not have imagined that the investigations and debate over Yucca Mountain would stretch on for decades, leaving nuclear waste to build up far beyond original expectations. The sparkling future of yesterday's science fiction has been replaced by a tarnished present and a dubious tomorrow. What Richter would have made of all of this, we can never know. He played a vanishingly small role in this larger drama, but the role he had to play, he played well. Some seismologists who had been only distantly aware of the Indian Point hearings were under the impression that it had not been Charles Richter's finest hour—that he had revealed a lack of expertise about earthquakes outside of his native California. In fact one is hard-pressed to point to any statement during his two-day testimony that has not stood the test of time. Sykes, meanwhile, looks back at Indian Point with a sense of frustration that the plant was not in the end built to withstand a larger earthquake. He also, however, expresses a measure of satisfaction that the seismology community has done its part to make sure that nuclear power plants are built to withstand future earthquakes. As citizens of the planet, we are left to hope that Richter's optimism for a successful nuclear future will also stand the test of time.

Supernova

It appears that my publicity quotient is as high as that of almost any one person at the institute, on a level with those who really have accomplished significant work scientifically. The comparison makes me blush, and I am sure it must grieve many of my colleagues who take me for a publicity-seeker.
—*Charles Richter, letter to Robert Sharp, May 29, 1964*

Once in a great while a magnificent light blazes brightly but briefly amid the familiar stars in the nighttime sky, a phenomenon known as a supernova. These spectacular events rank among the most energetic explosions that occur in the known universe; so brightly can they shine that they have been described for millennia before astronomers had the wherewithal to understand them. Supernovas represent one class of the so-called variable stars that Charlie Richter observed during his teen years in Southern California. These spectacular events are, however, rare in the earth's nighttime sky, occurring on average perhaps once per century in our own galaxy. They occur in other galaxies as well, of course, but at a very great distance from Earth.

It is thus virtually certain that young Charlie Richter never witnessed a supernova. Standing (and no-doubt feeling) alone on dark nights in the then-remote Los Angeles region, he could probably never have imagined that he would some day become one. In cosmic terms every human life shines brightly for the mere blink of an eyelash. What set Richter's life apart was not only the intensity with which it shone, but also the exceptionally brief duration of its greatest brilliance. Born with a unique combination of staggering intellectual gifts and equally staggering emotional and (probable) neurological problems, Richter spent most of the first three decades of his life struggling to come to grips with himself and to find his place in a world that usually left him feeling baffled and lost.

Richter's star blazed across the terrestrial landscape, especially within the halls of research science during the 1930s and into the 1940s. Colleagues saw a man who had overcome the debilitating emotional instability of his very early adulthood, although he remained rather peculiar and intensely private. In fact his demons never disappeared, or even retreated very far beneath the surface; they began to get the upper hand once again as he grew older and perhaps developed ancillary health problems that further affected his energy level and mental faculties.

In some fields of inquiry, particularly mathematics, scholars typically make their mark when they are very young. Pushing the edge of the envelope in highly abstract fields may require the neurological elasticity of youth. The earth sciences are different: they do not as a rule require pure abstraction but instead the ability to digest enormously complicated data and physical systems to arrive at the underlying order. At age forty-five, Richter had reached the time in his life when many earth scientists have achieved stature in the field and remain productive and creative. Instead he found himself unable to see projects through to completion without a stern taskmaster in the form of Beno Gutenberg. Although Richter maintained the highly tedious routine analysis necessary to continue the day-to-day operations of the Seismo Lab, and critical to the development of the magnitude scale, by his own account he became more and more dilatory in other aspects of his life.

His attentions wandered to other women; he wrote mountains of poetry as well as drafts of several manuscripts—all of them quite hopeless in their lack of focus—even as he moved into the role of elder statesman of seismology, a role to which substantial public recognition and admiration was attached. Among the ironies of Richter's life is that as an older man, he found himself on the receiving end of gushing attentions from women.

Richter's papers reveal incontrovertible evidence that he flirted (awkwardly) with female reporters and had several serious affairs of the heart (and body) in his later years. These relationships buoyed his spirits, as he expressed in some of the poetry written during this time. In his later years Richter's poetry became especially reflective. In the following untitled poem he returned to an issue with which he had grappled at some length as a younger man: the fate of the introverted soul versus those of individuals whom Richter perceived to be happy extroverts:

Count no man good, nor worth a penny,
Who makes his way by talk and noise.
Men sounding out like brass are many,
Or beating drums like little boys.

Prefer the quiet thief, the knave
Who breaks a path by fist and force,
To those who, neither wise nor brave,
Take shouting for their only course

Pity them later, when they go
Beyond this roaring life to face
Unscintillating stars, and know
The silence of unbounded space.

(Naturally extroverted, boisterous individuals may well take exception to the suggestion that only reserved introverts are truly brave and wise.)

Other poems penned during these years reveal similar themes, a growing preoccupation with mortality in particular. Another untitled poem, dated January 21, 1965, reveals a keen awareness of not only mortality but also of declining faculties:

What, over now? Does nothing yet remain?
Where there was such an unremitting flow
Of words and feeling, not so long ago
Written in hope, but inked across with pain.
This sheet I handle shows an ancient stain,
Erased in joy or fear; I hardly know
Why here a line was crossed and altered so—
I cannot raise the ghost to change again.
There's nothing half so strange as one's own self
When time has moved, and shifted all the stars,
And let the ocean rise upon the land;
A book with seven seals, laid on a shelf
Behind a screen of gold with iron bars,
Ciphered in words one cannot understand.

In a 1964 letter, Richter expressed similarly melancholy sentiments to former colleague Hugo Benioff, who had recently retired. The letter expressed the unmistakable warmth of kinship between old friends: "Every once in a while, " he wrote, "I come across a copy of the 1929 conference picture, and I am always shocked to see what kids we were then." He goes on to say, "The older I get, the more I appreciate Harry [Wood]. After all, without him we wouldn't be here. (I hear you say, "Is that good?") When he was wrong it took the entire cosmic process to stop him; but often he was very right. I think one of his most inspired remarks was 'Life is a husk.' "

The letter then veers into the autobiographical mingled with the retrospective: "Do you remember Harry's burglar alarm that got touched off accidentally now and then? We had four stations, and in the measuring room each one had four seismogram drawers . . ." The letter continues for a while in a similar vein:

> I always thought one of the best things you ever did was one of the first—developing a really accurate drive with Harry breathing down your neck—even if it did then take years to convince him that we were still throwing away data by using those confounded kitchen clocks to mark the records. If I remember right, it took Fred Henson's rather clumsy clock to prove it.
>
> Speaking of drives, one of the most powerful that got installed here was Beno Gutenberg's. I doubt if I would ever have accomplished much without his push; Harry was too slow. What with Gutenberg's arrival and your construction of the variable-reluctance seismometer we finally got off the ground, depression or no depression.

And then Richter veers towards the confessional: "Many things I no longer recall sharply." And, later, "About 1947 I felt as if I has [*sic*] barely waked up from a long and continuous nightmare; I found myself often thinking of happenings and publications ten years old as if they were just yesterday."

Here, however, the confession ends. He goes on to recall the 1952 Kern County earthquake and observes, "We need another badly, not necessarily that big. Can't you do something to speed up the strain now? I get tired of being quoted as saying that a big one is 'overdue'—what a word!" The letter ends with the same expressions of friendship with which it began: "you will be missed around here," Richter observes, before wishing him every happiness.

Richter retired from Caltech in 1970. In a one-page letter dated June 30 of that year he described the work that he would continue to do through his final week, after which he would leave for the mountains that had always been his sanctuary—his salvation. He gave an account of the status of his final tasks, and offered that he would always "be glad to give [his] opinion in difficult cases or check up on individual measurements." He then noted, "BG always did that!" He wrote that later that week he would prepare a memo for Jim Brune, who had been hired to take over the lab,

Fig. 19.1. Charles Richter (*center*), Lillian Richter (*right*), and Vi Taylor (*left, background*) at his retirement party in 1970. (Photo courtesy of Caltech Seismological Laboratory, reprinted with permission.)

and the staff to describe and explain the lab routine in detail. The final paragraphs appear to have been added later:

> If I can't get here tomorrow afternoon, will try to look in during the evening. In any case I expect to be here Tuesday a.m..
>
> It is now July 1, G.C.T. i quit.

The letter was signed, "Good hunting! CFR."

As we have heard, Richter's retirement party at the Seismo Lab spawned one bit of Richter lore that went on to infamy in seismological circles. One suspects, reading the letter that Richter sent later to colleague Bob Sharp, that it wouldn't have taken much at all to set him off that day. He was, after forty years, bidding a formal farewell to the only professional home he ever knew. It was clearly a reluctant farewell: he had for years resisted both Caltech's and his wife's urgings to retire. He might not have chosen seismology as a young man, but eventually he embraced it with a vengeance, and let go only with great reluctance.

Around the time of his retirement Richter mentioned in a number of letters his plan to spend his newfound free time finishing projects and writing articles, a promise that was at least partially made good. He wrote

Fig. 19.2. Charles Richter at his retirement party in 1970. (Photo courtesy of Caltech Seismological Laboratory, reprinted with permission.)

a number of papers in the closing years of the 1960s, including one study assessing the earthquake potential of a fault along the central California coast and one in *Science* that considered the relationship between earthquake locations and mapped faults in a number of locations around the world. In the late 1960s Richter had been one of the reviewers of one of the seminal plate tectonics papers: "Seismology and the New Global Tectonics," by Brian Isaacks, Jack Oliver, and Lynn Sykes. The theory of plate tectonics had not been developed based on seismological data: this paper was, as the title suggests, the first to integrate results from seismology into the new paradigm. Not every senior scientist at the time rushed to embrace the new ideas—including Harold Jeffreys, who, had it not been for Beno Gutenberg, might well have been Richter's close collaborator. Richter, however, wrote a very positive review of the paper and offered six pages of detailed constructive comments about matters such as the nature of seismic belts and faults worldwide.

Richter submitted his detailed remarks, without the positive overall assessment, as a formal comment on the publication, which was published along with a reply from the authors in the *Journal of Geophysical Research.* His colleagues read the comments as criticism, and, as had so often been the case throughout Richter's life, misinterpreted his intent, not realizing that Richter had embraced and applauded the paper as a seminal contribution.

Although Richter remained active on the seismology scene for many years following his retirement, by the time of the San Fernando earthquake eighteen months later he was already far removed from the day-to-day affairs of the Seismo Lab. His overall health and energy levels declined. In a letter to a colleague in March 1969 he wrote that the "accumulated fatigue of the day tends to disable me in the evening." Such accounts pop up in letters from other times as well, both before and after his retirement. In 1961 he mentioned in a letter to a Japanese colleague that he had "not been well, not really ill, either." He went on to write that he had been "able to go to the lab regularly, and something has been accomplished, especially minor routine work. Nevertheless I am continually fatigued, although the doctors seem to find nothing wrong." As early as 1955, he wrote in a letter that he had recovered from a bad nervous condition, although mentions of such personal matters are few and far between in his professional correspondence. For a man who would live to a respectably ripe old age, he clearly felt markedly unwell for many years.

It was, moreover, not only his opinion. As early as the mid-1950s, colleague and department head Bob Sharp observed that "Gutenberg is old, he's got heart trouble" and that "Richter isn't in much better shape, and Benioff is not a healthy fellow." Sharp's letter went on to express concern that the seismology program at Caltech could disappear in a heartbeat, or rather three heartbeats, and that the department should thus consider bringing a younger scientist on board. Eventually they did, in the person of Frank Press.

Both Sharp's and Richter's own accounts again suggests that he might have suffered from a prediabetic condition or perhaps hypothyroidism, either of which is typically associated with low energy levels and a decline in mental sharpness. His low cholesterol levels could have also had a deleterious effect on his mood as well as, potentially, overall brain function. One might suspect a more degenerative condition such as Alzheimer's, but, as we have seen, this diagnosis is belied by evidence that Richter retained possession of, at the very least, a healthy percentage of his marbles through his seventies. While he published no papers after 1969 except for a short letter to *Science*, his 1979 interview with Ann Schied reveals an impressive memory of past events as well as a marked lucidity of thought. Various letters through the 1970s—as late as the early 1980s, even—reveal similar qualities. Karen McNally, who worked with Richter from the time she arrived as a postdoc in 1976 until she left the Seismo Lab in 1982, deeply valued his insights and acumen. At the memorial service following his death she described the frustrations of his later years, when his mind remained quick but his body could no longer keep up. At times he would have to pause to regroup before continuing a conversation. "Getting old is such a damned nuisance," he told her.

Kate Hutton, now Caltech's lead staff seismologist of the Southern California Seismic Network, arrived at the Seismo Lab in 1977. She landed in earth sciences much as Richter had a half-century earlier, by way of an earlier passion for the stars. Hutton earned her Ph.D. in astronomy and worked as a postdoc at Goddard Space Center before finding her way back down to earth. Richter no longer came to the lab regularly by the late 1970s, but he remained a regular visitor after newsworthy earthquakes around the world and close to home, including the 1979 Imperial Valley earthquake, a series of earthquakes and unrest near Mammoth Lakes in 1980, and even a relatively modest magnitude 5.5 quake offshore of Santa Barbara in 1978. By this time Hutton was on her way to becoming the lab's most visible scientist, a part of the fabric of life in Southern California. (A female seismologist in the 1997 movie *Volcano* was widely taken to be Hutton's alter ego. She expressed a twinge of dismay that "her" character perished in rather gruesome fashion during the film.) Richter gave no interviews during these years; instead he mostly poked around, examining seismograms and learning—for nothing more than his own satisfaction, it seems. He seemed happy to be back in his element after big earthquakes: "He used to whistle Souza," Hutton recalls.

Like McNally, Hutton remembers Richter fondly. Although she never knew him well, she describes the day that he wandered into her office, having apparently found out that she held a position of some authority at the lab, "plunked himself down," and began talking to her (one should perhaps say "talking *at* her") about matters related to network operations, including his thoughts about some of the original network stations. A young scientist just barely beyond her postdoc years, Hutton found herself amazed to be the one-woman audience while the world-famous Charles Richter held court. Echoing the experiences of women like Vi Taylor and Betty Shor, Hutton never had the sense that it mattered to Richter that she

Fig. 19.3. Kate Hutton. (Photo courtesy of Caltech Seismological Laboratory, reprinted with permission.)

happened to be a female seismologist—even at that time a rare breed. Hutton's own experiences with the news media left her with further admiration for Richter's efforts to communicate to the public in language that people could understand.

By the end of his eighth decade, however, Richter's universe began to contract. He visited the lab less and less; his long hikes in the mountains were replaced by much shorter walks closer to home. Never a social butterfly, Richter was not well known among his neighbors at any time in his life, but Lawrence Fusha, who lived a few houses down from the Richters on Villa Zanita, recalls his famous neighbor with clear affection. He describes a day after the 1971 San Fernando earthquake when the Richters extended a rare invitation to his neighbors to come over for coffee and cookies and a look at a seismogram of the earthquake recorded on the seismometer still operating faithfully in the Richters' living room. Fusha also recalls with appreciation the world-renowned scientist's cheerful willingness to help with his stepson's school science project. Richter did become somewhat more social with his neighbors in the aftermath of Lillian's death: he began to wander round in the evenings, knocking on doors just to chat. "He seemed despondent," Fusha recalls. The partnership might have had more than its share of complexities, yet Charles and Lillian's union had endured—perhaps in some respects a marriage of convenience, but in the end so much more than that.

Two colleagues knew Richter especially well during the last decade or so of his life: colleague Karen McNally and business partner Eric Lindvall. Richter remained active in the business through the 1970s and early 1980s. Lindvall remembers his former partner and friend with open admiration and affection. He describes his interactions with Richter as a "great learning experience." Lindvall witnessed and heard his share of "absentminded professor" stories, like the time Richter dressed for a black-tie occasion and forgot the tie. He also recalled Richter's legendary temper.

Still, mostly Lindvall recalls the Charlie Richter that his closest earlier colleagues had known and "sort of loved" for forty years: a man of tremendous scientific insight and dedication; a man who minced no words, ever. In his capacity as a consultant Richter was called on to review reports and to contribute to hazard assessments, including various projects of the Department of Water and Power. Always the scientist, never the diplomat, Richter answered the questions that were put before him. When the Torrance City Council asked Richter about the possibility that local oil drilling might trigger earthquakes, Richter replied that such operations were restricted to the very upper layers of the crust and were not likely to influence earthquake activity. A councilman pressed further: "What was the *probability* that something might happen?" Richter did not answer for a few seconds, and then a few seconds grew into a very long and quiet minute, maybe two. Finally, amid the increasingly awkward hush, Richter responded, "You just hit on one of my pet peeves. If I told you a number you wouldn't know what to do with it." After the few seconds of further silence that greeted this response, the council chairman asked weakly, "Any more questions for Dr. Richter?"

Lindvall further recalled Richter's enduring skepticism about the feasibility of earthquake prediction, and of efforts of close colleagues to use it to sell an earthquake research program. As chapter 16 discussed, these efforts paid off with the enactment of the National Earthquake Hazards Reduction Program in 1977, a program that has supported a tremendous amount of useful research that has had nothing to do with prediction—as well as some research that has addressed the question of prediction.

In a recorded oral history in 1991, Clarence Allen described the genuine optimism that many top scientists had felt in the 1970s, that earthquake prediction might soon be possible. In retrospect, Allen went on to add, "We would have to say that our hopes that we were going to have meaningful, short-term predictions of earthquakes within a ten-year period were optimistic and naive. That has been disappointing." A pragmatic scientist like Lindvall—or Allen—can look back with a measure of equanimity: the pursuit of earthquake prediction had been well intentioned, not a "carrot" but rather the basis for a sensible research program motivated by promising developments. And as Allen said in a 2002 interview, "The excitement was over prediction, but I think many of us had in the back of our minds the hope of getting better quantitative understanding of the probabilities of future earthquakes, not just short-term predictions." The pragmatic scientist can also look back with an awareness that earthquake prediction helped launch a valuable research program, one whose other merits were not easily appreciated by politicians or the public. For Charles Richter, who often did not seem have a pragmatic bone in his body, the hopes for prediction had been naive and misguided from the outset, and ends did not ever justify means. By some accounts his unyielding views on the subject of earthquake prediction drove a wedge between Richter and his closest Caltech colleagues. Even Allen took exception to his colleague and longtime friend's intimation that earthquake prediction was—and, perhaps, would always be—a business for of charlatans: "I do not," he said, "think that Charlie's statement demonstrated much vision for the future."

Towards the end of his career Richter also felt unease and regret over the extent to which defense contracts were supporting Seismo Lab research. Since the inception of the World-Wide Standardized Seismographic Network in 1961, the Department of Defense had continued to provide substantial funding of seismological research, much of it focused on the seismic waves generated by nuclear tests. Many seismologists put in their bids for defense support: in many cases such grants allow them to also investigate scientifically interesting problems. Here again Richter lacked the capacity for compromise that allowed many of his colleagues to make their way in an increasingly complex world. In his black-and-white world, interesting defense-oriented research was a shade of gray that didn't exist—that *shouldn't* exist. An academic department should not be a DOD shop.

One has to wonder, what other wedges had been driven? At the Seismo Lab today one finds the Benioff Conference Room and the Gutenberg Library, but no room or facility bears the name of the scientist who remains, even in death, the lab's brightest luminary in the public arena. Among seismologists, especially at Caltech but also elsewhere, there remains the enduring consensus that, of course, colleague Beno Gutenberg had the stronger legacy as a *scientist.* The rest of the sentence is unspoken: Richter achieved greater fame in the public arena because of his penchant for "grandstanding" and the public's lack of understanding of science. Ever the keen observationalist regarding himself as well as earthquakes, it seems Richter knew from whence he spoke in his 1964 letter to Bob Sharp: "I am sure it must grieve many of my colleagues who take me for a publicity-seeker."

There is in fact a long-standing disconnect in science: many of the scientists who achieve the greatest fame in a public arena are far less venerated—if not outright vilified—within their own communities. For scientists whose lives revolve around the sometimes arcane and always conceptually difficult issues that define the edge of the envelope in a given field at a given time, communicating science to the public can be seen as a rightful task for lesser mortals. Carl Sagan, Steven Jay Gould, Charles Richter—none were nearly as respected by their colleagues as by the public. A disconnect between public fame and scientific respect is sometimes warranted: an especially telegenic scientist is not necessarily an especially accomplished scientist. Richter was not terribly telegenic; one doubts he would have achieved the same fame had more of his life overlapped with our modern television era. However, in his own inimitable way he was a gifted communicator as well as a highly accomplished scientist.

It would not be entirely apt to describe Richter as a prophet without honor in his own country. It would, however, be fair to say that Richter's honor within his own country did not match that of his honor outside his own academic community. His peers never elected him to the National Academy of Sciences; many still point to his encyclopedic textbook rather than the magnitude scale as his most important contribution. Some seismologists go so far as to say that Richter was really little more than Beno Gutenberg's assistant. Lindvall expresses regret at this, as do many of the scientists who had been in positions to appreciate fully the depth and breadth of Richter's talents and contributions.

Richter once observed that seismology owed an enormous unpaid debt of gratitude to Harry Wood. One wonders if it doesn't also, ironically, owe another to its most famous founding father. Richter's public fame aside, what remains are groundbreaking contributions in observational seismology; a book—by some accounts still without peer—that is nothing less than an "encyclopedia of earthquakes"; an unflinching, uncompromising integrity about key issues in earthquake hazard as well as earthquake prediction. In developing the magnitude scale he did what most needed to be done at the time, and he did it brilliantly: he laid the foundation necessary for seismology to move forward as a quantitative, rigorous science.

Both Richter's landmark 1935 paper and a number of later writings, moreover, reveal that he anticipated several key developments that would not be recognized by the larger seismological community for years, even decades, including the increase of moderate regional earthquakes in the decades preceding a major earthquake and the existence of remotely triggered earthquakes. Without question, he did not push the field of seismology forward with the relentless drive of his longtime collaborator Beno Gutenberg. But Richter had other gifts. He didn't answer the next burning question, and the next after that; he was the type of scientist who sometimes understood answers to questions his colleagues hadn't asked yet. Although he did not write a number of papers that he might have written—papers that easily could have been landmark contributions—his "citation index," one barometer by which scientific accomplishment is measured, exceeds that of Gutenberg. Gutenberg's best-cited research papers, meanwhile, are those he coauthored with Richter.

Of course, Beno Gutenberg was the more accomplished scientist. Seismologist Leon Knopoff, who spent a year at Caltech in 1961 before leaving for UCLA, describes Richter as a "collector of facts about earthquakes," a less insightful scientist than either Gutenberg or Hugo Benioff (the latter a towering figure in earthquake science, his modest role in this story notwithstanding).

Beno Gutenberg made numerous seminal contributions to earthquake science; Charles Richter became famous. No single easy answer explains the paradox, but certainly part of the answer is easy: the public came to know Charles Richter because he talked to the public. In the aftermath of the 1933 Long Beach earthquake an area resident, H. L. Barlow, assembled a forty-page scrapbook of newspaper articles that eventually found its way to the manuscript collection at the Huntington Library. In the relatively few articles that deal with scientific matters, one encounters three names: renowned Stanford geologist Bailey Willis, Seismo Lab director Harry Wood, and Charles Richter. Early Caltech faculty members such as Gutenberg and John Buwalda were nowhere to be seen. Willis, already in his seventies in 1933, died in 1949. Wood would have been in his fifties at the time, laboring under serious health problems; he died in 1958 but had disappeared from the public scene even earlier. That left Charles Richter, just shy of his thrity-third birthday, with many more decades ahead of him as a scientist and, by default, Caltech's leading earthquake spokesman. It was a role he continued to play for decades. When Barry Keller, who went on to a career in hydrology, arrived at Caltech as an undergraduate in the mid-1960s he knew one and only one Caltech professor by name: even as a teenager growing up in California he knew Richter's name—he had seen it in the newspapers.

Following Richter's death, *Pasadena Star News* editor Wanda Tucker wrote an article recalling a man she had come to regard as a "down-to-earth friend." She described his patience; the fact that he had never sounded too busy to accept calls. "Believe me," she wrote, "that really made him special to all of us in the news business who over the years needed information on various topics." Her appreciation beams through her words: "He liked and respected newspaper people, as we liked and respected him." She further applauded his talent at talking, "in terms the average reporter—and the average reader in turn—could understand."

In this regard Richter stood in contrast to many if not most of his colleagues, whose views about journalists are manifest in comments such as that by Beno Gutenberg in 1958: "I have no doubt that what happens when a layman unfamiliar with the subject, writes [a scientific article], there is always some nonsense in it."

In addition to Richter's talent for—and interest in—communicating science to a broad audience, one of the ironies of Richter's life was that, notwithstanding an ambivalence towards a field he never really chose, he was in the right place at the right time. As one colleague told Kate Holliday in 1971, "The seed fell on the right ground at the right time." Through the first half of the twentieth century, seismology was what scientists would regard as largely *phenomenological* science, which is to say, it was more concerned with basic descriptions and quantification of earthquake phenomena than with a true understanding of the physics of earthquakes. When scientists speak of phenomenological science, they mean science that describes more than it explains. The latter is considered more important, but still today, invariably follows the former in earthquake science. Given the enormous complexity of earthquake processes, questions such as "What causes aftershocks?" cannot possibly be addressed without a substantial arsenal of observational results describing the nature and statistics of aftershock sequences. More fundamentally, one cannot describe the statistics of aftershocks until one has developed the tools to size up earthquakes. Richter recognized that the development of such tools did not equate to the kind of fundamental scientific contributions that he had yearned as a young man to make. Yet the field of seismology could not move forward without an understanding of basic earthquake phenomena. As seismologist Freeman Gilbert wrote after Richter's death, *Elementary Seismology* was "really not so elementary": it was a key part of early efforts to establish seismology as a quantitative science. Kei Aki described Richter as a "keen observer, a man of letters, an excellent linguist."

Ironically, Richter could not have been better equipped to make such contributions. He not only became a walking encyclopedia of earthquakes his unique cognitive makeup endowed him with the tireless tenacity needed to distill solid phenomenological results with only limited data and even more limited computational tools at his disposal. In combination with his remarkable scientific acumen, this made for a remarkable scientist.

Richter's later scientific contributions did not match those of his earlier years, but the same can be said for many scientists. Given the evidence that Richter was probably plagued by physical as well as neurobiological difficulties throughout his life, one is inclined to be impressed that he did as well as he did for as long as he did. A baby born in the year 1900 had a life expectancy of forty-eight years. This number reflected the very high infant and childhood mortality from infectious diseases in the years before antibiotics were discovered; having been born in 1900 and survived to adulthood, by 1950 Richter's life expectancy, in actuarial terms, had increased to about seventy-two years. By any standard Richter beat the actuarial odds—perhaps against all odds. His longevity might have been due in part to the fact that he remained quite active throughout his life. Inevitably, however, he could not beat the odds forever.

Eventually time began to catch up to Charles Richter. Apart from his own account of health issues, his correspondence in the late 1970s reveals a decided turn towards the eccentric (or rather, towards the even more eccentric). In 1978 he wrote a local gossipy newspaper column to note that a sign on the 210 freeway gave mileage to La Cañada with the tilde mistakenly placed over the second *a* instead of the *n*. "Perhaps," he observed, it had been "put there by some very young man with a misplaced eyebrow." He began to write others sorts of letters to newspaper columns and others. In 1980 he wrote a sweet but arguably mildly eccentric letter to the proprietors of the "Town House for Dogs," to express appreciation for the care they sometimes gave his canine companion Jack. These correspondences were not the ravings of a lunatic; indeed, Richter remained lucid in interviews and some letters as least as late as 1982. The other sorts of letters were, however, the writings of a genius whose faculties had begun, inexorably, to slip. His journal entries grew more and more ragged.

Yet as his private star began the inexorable process of dimming, his public star shined all the more brightly. Any earthquake as large and destructive as the San Fernando temblor of 1971 will kindle newfound interest in seismology amongst the public. Richter found himself in the elder statesman role, called on to offer his opinion to officials, the media, and the public. He received scores of letters throughout the 1970s: formal requests for portraits to be included in exhibitions, informal requests for information and, especially, autographs. By Clarence Allen's account, Richter did enjoy the attention from the media and public, which came at a time when he was receiving rather less approbation from his colleagues.

Richter was not without honor in his own country, even though some colleagues resented his public fame. In 1976 Richter received the Medal of the Seismological Society of America, the highest honor bestowed by the preeminent society of research seismologists in the United States. In 2004 the society honored Richter further by creating the Charles F. Richter Early Career Award. The invitation list to his eightieth birthday party held at the Seismo Lab reads like a who's who of American seismology in the late twentieth century: not only old friends and Seismo Lab colleagues like Vi Taylor and David Johnson (and some widows), but also a few of the brightest young stars in the field. Nobody sang any clever songs at this party.

Reporter John Barbour from the *Bremerton Sun* caught up to Richter in early 1981 to interview him for articles that he published in the *Los Angeles Times* and *San Gabriel Valley Tribune.* At this time, going on eighty-one,

Richter was still living alone. Barbour described the small cottage in Altadena: cluttered with books, magazines, and mail; the seismograph that remained in the living room, its pen long ago stilled. During the final years of his life Richter learned about large earthquakes the same way that everyone else did: from the news media. Just a few months shy of his eighty-first birthday Richter remained a walking encyclopedia of earthquake information. His memories were sharp enough to relay a story from nearly a half-century earlier, about finally returning home in the wee hours of the morning after the Long Beach earthquake struck near six o'clock in the evening on March 10, 1933. By Richter's account, Lillian had been "a California girl, accustomed to earthquakes." Their cat at the time felt differently, however: it was "very displeased and spat at the floor because it wasn't behaving properly."

The man who had not been safe in a chemistry lab at age twenty or with a camping stove at age fifty was at eighty-one living alone, shopping for his own groceries, and cooking at least some of his own meals. (His journals during his sixties and early seventies reveal that he ate many of his meals, including many breakfasts, in restaurants: presumably this habit continued during his later years well.) His one constant companion, Jack, was described in a 1980 *Los Angeles Times* article as being of "doubtful ancestry but loving disposition"; Barbour described the animal as "a kind of genetic clutter." Richter had not adopted the mutt, exactly; rather the dog had "wandered in one day and adopted [him]." By his later years Richter employed a weekly housekeeper who maintained some semblance of order in the place. Or rather, she tried to maintain some semblance of order. Richter told John Barbour that he tried to tidy up so that the housekeeper could clean; upon hearing this, Eric Lindvall laughed and observed, "He didn't try very hard."

Richter continued to manage his own affairs as well, during his later years with help from longtime friends Jerene and Bill Hewitt. Jerene and Bill were among Richter's "writer friends": Jerene had been part of the book club that the Richters had belonged to since the 1940s. She taught in the English department at Pasadena City College during the 1960s and early 1970s, and served as faculty advisor for and sometimes contributor to the literary magazine, *The Pipes of Pan* (later, *Inscape*). Former colleagues describe her as brilliant, an accomplished poet as well as teacher; her name (Jerene Cline Hewitt) appears in several *Who's Who in the West* volumes in the 1970s and 1980s. She died in 1998. Karen McNally describes Hewitt as a "classy woman," one who came to feel a tremendous sense of responsibility for Richter's care.

As is often the case in modern times, the chore of driving an automobile began to loom as the biggest problem for a man of advancing years and declining health, even if friends and loved ones can provide assistance with other day-to-day affairs. Never the best of drivers even in younger days, by his later years Richter's car-related mishaps began to involve more serious incidents than denting the fins of his 1957 Chevy backing into the eucalyptus tree next to the driveway of his Altadena home. Some of his difficulties were benign, involving only stationary automobiles. In February 1980 he wrote a gracious letter to Pasadena police chief McGowan, expressing his appreciation towards a courteous officer who "did not even laugh" when Richter showed up one day concerned that his car had been stolen, when in fact he had parked it near downtown Pasadena and forgotten where he left it. He lost his car on other occasions as well in early 1981.

A more serious automobile-related incident took place a few days after Christmas of 1981, when Richter and the car he was driving careened off the road into a ditch near the Jet Propulsion Laboratory. He wandered in a daze for six hours before being rescued and sent to a local hospital, where he remained for two weeks. The incident marked the beginning of the end of Richter's life as a man of science and a public figure. From early 1982 onward he had daytime nursing care at his house, much of it provided by longtime friends rather than paid help. Friends arranged for meal delivery to his home from Meals on Wheels.

In the fall of 1982 he wrote to the then president of UC Santa Cruz to express his gratitude for and acceptance of their proposal to name their new Seismological Laboratory in his honor. This had been arranged by Karen McNally, who left Caltech to become a professor and then the director of the seismological laboratory in Santa Cruz. Richter wrote that his health had improved following two hospitalizations. "Everyone remarks on my recovered appearance," he added. He noted that medications still restricted his mobility, but that he would be able to make the trip in the company of a "good friend and business associate." He did, in fact, attend the dedication, in the company of Eric Lindvall, in March 1983. He was once again in the company of luminaries: McNally, Hiroo Kanamori from Caltech, Bruce Bolt from UC Berkeley, and others. It was the last big public event of his life.

Richter pulled away from the public arena as he approached his mid-eighties. As his health declined he continued to live at home, but with continued nursing assistance. Friends rallied on the occasion of his eighty-fourth birthday: he received dozens of cards from former colleagues and friends from all walks of life (all of *his* walks of life). These cards stand in testimony to the remarkable life of a man for whom interpersonal relationships had never been easy. Throughout his lifetime he found ways to reach out, to establish the connections that he longed for since childhood but struggled as a young man to find. Those who had gotten to know him—some through the Glassey group, some through the book club that he and Lillian attended for decades—came to cherish his friendship.

A reported heart seizure in July 1984 landed Richter in the hospital; by August he had been moved to a convalescent hospital. The August 21, 1984, *Los Angeles Times* described the "earthquake scale inventor" as in failing health, near death. In fact he lived another year, but never regained his health. He died on the afternoon of September 30, 1985, of congestive heart failure, at the Park Merino Convalescent Center in Pasadena. His cremated remains were buried in the same gravesite as Lillian's remains, at Mountain View cemetery in Altadena. For years after Richter's death he remained an anonymous resident of Mountain View: nobody bothered to replace Lillian's original marker. In 1996 a private donor contributed a new marker, and a small group gathered for a ceremony: a couple of Richter's colleagues as well as a dozen or so other friends. To the left of the Richters' gravesite lies another poignant headstone: Reginald F. Saunders Jr, 1925–1957. Apart from their proximity, nothing on either headstone provides any clue that these three individuals were bound together in life as well as in death.

The headstone to the immediate right of the Richters' reads "IN LOVING MEMORY, MARGUERITE BARSOT, 1900–1991." Although it is not clear how they met, in 1920 Marguerite and Lillian were two of six people living together in a boarding house in Los Angeles. If they hadn't already known each other when they moved into the house, as the only two young women living with four men ranging in age from twenty-three to fifty-one, they would have felt a natural kinship. In fact they became more than just friends, they became best friends, and best friends they remained for life. The nexus between the poignant trio of grave markers is thus not Richter himself but rather Lillian: she was buried next to her only son, joined later in death by her husband and the dearest female friend of her adult life. But Richter is with her at the heart of this intimate galaxy, his presence properly, if belatedly, acknowledged: forevermore part of a family that might have broken the standard nuclear mold with a vengeance and yet, in its own way, endured.

On the day in April that I first visited the Mountain View cemetery, the Southern California sun shone brightly over the gently rolling green lawns. Birds sang in the trees, squirrels scampered across the grounds, a pair of majestic red-tailed hawks soared just overhead—an oasis of solitude in

Fig. 19.4. Grave marker for shared gravesite of Lillian and Charles Richter. In 1996 this marker replaced an earlier one that listed only Lillian's name. (Photo by author.)

verdant hues. The clatter, crash, and bang have been silenced at last. Charles Frances Richter now rests in peace, not quite in the mountains that he so dearly loved, but right at their doorstep. It is only fitting that Charlie Richter has become, in a literal as well as a figurative sense, a part of the Southern California landscape.

The date of Richter's death—September 30, 1985–may ring a bell to earthquake cognoscenti: Richter died less than two weeks after the devastating magnitude 8.0 earthquake in the state of Guerrero, Mexico. Although this earthquake occurred along the west coast of the country, the lion's share of the damage and death toll (nearly ten thousand lives) were concentrated in Mexico City, some 360 kilometers from where the earthquake had struck. The staggering loss of life was almost entirely due to poor building design and construction, in particular of high-rise buildings set into motion by amplified shaking within the ancient lake-bed zone beneath the city.

In his final days, Richter saw images from this disaster on the television set in his hospital room. He was not always aware of his surroundings, but to those around him he seemed aware of the events taking place several thousand miles south. What went through his mind, one cannot begin to imagine. Virtually every seismologist feels a crushing sense of impotence when an earthquake disaster plays out—nowadays, on televisions for all the world to see. Seismologists today understand what Richter understood half a century ago: earthquakes exact a tremendous toll, and this price need not be paid. Seismologists dedicate their lives to a science with direct, substantial societal impact. A better understanding of earthquakes leads to a better understanding of earthquake hazard, and in turn to hazard mitigation.

To the general public, Richter's immortality has everything to do with the scale that bears his name. Yet Richter's legacy is far more profound and enduring than this one contribution, important though it may have been in its day. In the words of colleague Thomas Hanks, "I realize that there were seismology and seismologists—even earthquakes!—before Charles and his colleagues, but the small group of people active in the 1920s and 1930s gets the credit, I think, for putting seismology on the map as a real scientific discipline." In so doing, this group led the scientific charge that in turn led directly to serious hazard mitigation efforts in California. In a very real sense, the state today owes a debt of gratitude to Richter and his colleagues for their relentless and exhaustive efforts to understand earthquake hazard (without sensationalizing it) and communicate their understanding to the public and policymakers. In his willingness to communicate at length with the press, Richter was nearly without peer. And if it was by virtue of a neurological disorder that he was such a "walking encyclopedia," we are left wondering where we would be if Charles Richter's brain had been wired according to standard blueprints.

As one nears the end of the story, one question remains difficult to answer: to what extent did Richter crave and seek out attention and approbation from the public and media, and to what extent did it follow from his abiding desire to impart his knowledge about earthquakes? That he might have enjoyed this approbation towards the end of his life does not imply that he actively sought it out—or, in particular, that he did so at the expense of others. We can consider his own words on the matter: they unerringly suggest that he was not, in fact, guilty of "grandstanding." A similar conclusion can perhaps be reached when one considers Lawrence Fusha's account of a neighbor who went out of his way to reach out to neighbors in the aftermath of a large earthquake. In Eric Lindvall's words, "He loved to teach people about earthquakes." By temperament as well as neurology Richter was never a comfortable lecturer in a traditional classroom setting or in front of any large audience. He was, however, in his element when he could hold forth on his favorite subject one-on-one: reminiscently in the form of the pedantic, one-sided (yet highly educational) conversations that are a key hallmark of Asperger's syndrome. Talking with reporters may have provided Charles Richter with his only comfortable outlet for the teaching that he yearned to do.

At the same time, one has to admit: that a person bemoans the "terrible burden" of media attention does not mean they don't seek it out at every opportunity. But all objective evidence suggests otherwise. He clearly did love to teach; to impart some of his encyclopedic knowledge to not only fellow scientists but also the public, for whom earthquake science was tremendously relevant. Another fact remains. Richter would be asked many times during his lifetime about Gutenberg's contributions to the original development of the scale: his answer never varied. Charles Richter was not *only devoid of pragmatic bones; he also, it seems, had nary a political one* in his five-foot, eight-and-one-half-inch body. When someone asked him a question, he answered it, truthfully and to the best of his ability. These answers never varied, not over the span of many decades. He had none of the politician's faculty for the savvy answer; none of a diplomat's talent for the artful reply. He gave credit where credit was due, no more and no less.

Richter's legacy endures. He and his colleagues, particularly Wood, Gutenberg, and Benioff, established a tradition of excellence in observational seismology research that continues. Some of Richter's colleagues are still at the Seismo Lab. They are joined now by younger seismologists, most of whom never met Richter, all of whom continue in his footsteps. The seismic network that Richter and his colleagues established has evolved considerably, of course, and is now run jointly by Caltech and the U.S. Geological Survey; it continues to generate the state-of-the-art earthquake recordings that drives state-of-the-art earthquake science. Even before the Seismo Lab became part of Caltech, the Southern California Seismic Network was a world-class research resource, the value of which Millikan clearly recognized when he brought in world-class seismology expertise in the form of Beno Gutenberg to more fully exploit data from the nascent network.

Richter's legacy also continues in the critical day-to-day business of earthquake monitoring and, in particular, earthquake catalogs. For a great many purposes in earthquake research and hazards assessment, consistent, accurate long-term catalogs of earthquakes are absolutely critical, but surprisingly hard to come by. Seismometers did not exist until the closing years of the nineteenth century, and early instruments were neither sophisticated nor numerous enough to provide consistent information about any but the largest earthquakes worldwide. The small group of people active in the 1920s and 1930s, at Caltech and a handful of other institutions worldwide, created the concept of modern network seismology. Wood, Anderson, and Benioff played key roles in developing reliable, portable, fieldworthy instruments. Richter was no mechanical genius, but via his computational insights and exhaustive, relentless assault on the data, he became a cornerstone of the earthquake monitoring and cataloging efforts in Southern California. This legacy remains very much alive. The Southern California earthquake catalog, which dates back to 1932–the first year that Richter began to make the measurements needed for magnitude estimates—remains one of the longest and best catalogs of earthquakes available for any region in the world.

Richter's star may have faded gradually, but in fact it shone with greatest intensity for only a brief duration. The work for which he is famous was to a large extent done within the span of a single decade. And, like a supernova, the light from this explosion continues onward to shine in ever more distant skies. Seismologists sometimes feel a sense of frustration, explaining how and why the "Richter scale" is now obsolete. In the back of their minds some seismologists feel an additional measure of frustration, thinking if not saying that it should never have been known as simply the Richter scale.

Yet the initial, groundbreaking, achievement was, at the end of the day, Richter's achievement. It is easy to fault Richter's sense of ownership of—and pride in—the contribution, yet by all objective evidence this sense was deserved. He built on earlier work by Wadati, but virtually all scientific contributions build on earlier work by others. He benefited from advice from colleagues Harry Wood and Beno Gutenberg, but virtually all scientists benefit from advice from colleagues. Gutenberg made important contributions to later extensions of the scale, but virtually all good research is expanded upon by others; a close colleague further described Richter as having been a full partner in these later efforts. As for the confusion over different magnitude scales, seismologists might be frustrated when the media and public continue to get it wrong, but maybe it is we who have missed the point. All later and better magnitude scales trace their lineage directly to Richter, but they do not bear his name. So familiar are we with magnitude values—the iconic magnitude 8 earthquake, the lightly felt magnitude 3 shock—that it is all too easy to forget: these numbers only have meaning because Richter gave them meaning in the first place. The very word *magnitude* has meaning in seismology only because Richter introduced it, drawing on the suggestion of a colleague and friend as well as his lifelong passion for the heavens above. When all is said and done, the Richter scale is Charles Richter's scale: measure of an earthquake, measure of a most extraordinary man.

APPENDIX

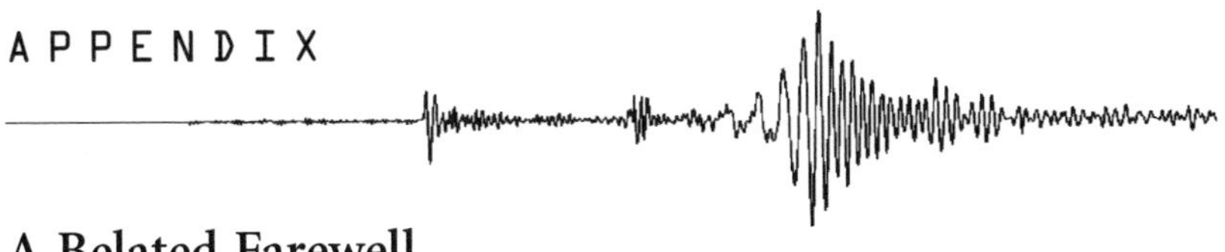

A Belated Farewell

> My main point today is that usually one gets what one expects, but very rarely in the way one expected it.
> —*Charles Frances Richter, 1970*

FOLLOWING a long and storied career, Richter never had the chance that he had wanted to offer a few deeply personal words on the otherwise unhappy occasion of his retirement. His retirement *party* was for Richter even more unhappy than the occasion that inspired it: it was precisely the kind of event that Richter had taken pains to avoid throughout his career. Talking with those who knew Richter during his long career with the Seismo Lab, one gets the idea that his speech would have been something of a bombshell: a thoughtful, extended philosophical statement from a man who had rarely shared anything of a personal nature with his colleagues.

The main point that Richter hoped to make emerges as remarkably prescient in retrospect: "usually one gets what one expects, but very rarely in the way one expected it." Richter did not get the chance to share his sage wisdom with the younger attendees of the party, but he gets a chance within these pages to share it with the world. Sadly, some of his thoughts on society and warfare ring as true to modern ears as they would have to the ears of 1970. And so his thoughts are in some respects no less timely for having been shared several decades later than he had planned. The draft of the speech ends abruptly. It appears he had more to say on that auspicious occasion: what, we will never know.

> *May 22, 1970*
>
> Those of you who know me best [know] that this occasion is not of my choosing. About this time I would have preferred to crawl into a hole and pull it in after me. I am grateful for your kind intentions; but no matter how you sugarcoat the pill, it still tastes bad.

The situation reminds me of an Irish wake; but on that occasion the dear departed is not expected to participate in the festivities; if he does it creates one of those weird situations familiar in folklore and legend.

Since I cannot join you in gay rejoicing, I shall ask to be heard in a few heartfelt remarks.

My main point today is that usually one gets what one expects, but very rarely in the way one expected it. That may sound cryptic; to make it clearer, let me go back a generation and quote from Tennyson,

> For I dipped into the future, far as human eye could see,
> Saw the vision of the world and all the wonder that would be:
> Saw the heavens filled with commerce, argosies of magic sails,
> Pilots of the purple twilight, dropping down with costly bales;
> Heard the heavens filled with shouting, and there rained a ghastly dew
> From the nations' airy navies grappling in the central blue,
> Far along the world-wide whisper of the south wind rushing warm,
> With the standards of the peoples plunging through the thunderstorms—
> Till the war drums throbbed no longer, and the battle flags were furled
> In the Parliament of Man, the Federation of the World.

Nowadays one wonders why Tennyson thought that wonderful. The realities of air freight, air warfare, and the United Nations are not what we rejoice most over.

My own experience has been a somewhat different song to the same tune. As a student I was much interested in the attractive possibilities of physical and biological science that seemed then to be in the near future—such as the exploration of space; the unraveling of atomic structure and the rationalization of chemistry; the release of nuclear energy; the development of what we now call molecular biology.

All those things came about, but with differences from expectation—and in a social environment vastly different from what we had hoped for.

I was impressed by the hopeful writings of H. G. Wells. He had a sound background in science; I could understand this thinking. He had a clear vision of the coming age of nuclear energy. Wells died a sour and disappointed man; yet he himself had written that the social advances of which he was the prophet might take centuries.

I need not tell you that right now there are all the necessary means to create a decent world. The chief obstacles are ignorance; greed; militarism; nationalism; and the violence that stems either from a psychotic impulse to destruction, or from a feeling of inferiority and a desire for revenge.

I have a sense of humor; but over the years that sense has developed one blind spot. I can no longer laugh at ignorance or stupidity. Those are our chief enemies, and it is dangerous to make fun of them.

After the two great wars there were great outbursts of hope for society. Both were disappointed. Those to whom the first is not a memory can find it described in Wells' "Outline of History."

Lately I have been shocked repeatedly by those who insist that we who lived through those two catastrophes should therefore be sympathetic to the officially sponsored side of the conflicts that are now going on. I feel that I have learned better than that. It is painful to hear the same arguments now that were used to justify the two great wars—and along with them the same half truths, and probably some of the same lies.

Naturally, young people notice some of this. They are restless and angry because they know they are being lied to, even though they cannot get at the truth. Some of these overlook that all the lying is not being done by one side; there is plenty of outrageous lying put forward by so-called revolutionaries.

In spite of the greatest development of information services the world has ever known, I think that no ordinary citizen has any real access to the facts on which he might base sound judgement on the national and social issues of our times.

I dislike many things that Spiro Agnew has said; but some of his remarks about the news media were justified and deserved. I speak from personal experience, over many years, with the manner in which news is handled.

A few remarks about the future. In spite of the present gloominess of the world outlook, I hope that the future of Caltech will be good. I do not enjoy some of the changes that are going on, but I expect to be round for a little while yet, and am willing to put up with the inevitable.

The Institute has chosen a very stormy time to broaden itself into a university, with the risk of bringing on to the campus more of the personalities and circumstances that have proved to be disruptive elsewhere. I hope that the event will justify taking this risk.

I do hope there will not be too much of the timidity exemplified by giving up the old Institute seal because it showed young people with a torch—or by dropping the worth[y] motto which accompanied the seal. "The truth shall make you free," One wonders whether this implies a less of faith in the great personage to whom those words are attributed.

I hope that the Seismological Laboratory will pro[s]per, and keep up its close connection with the immediate problems of local earthquakes. I regret that several good men, staff and students, who were interested in these problems, have left us.

I recall that many years ago, when much of my time was going into research with Gutenberg on seismic waves and the interior of the earth, someone repeated to me a comment by an engineer: "Oh, those fellows aren't interested in earthquakes." From the engineering point of view, I felt that this was partly justified, and I took it as an incentive to get closer to the immediate facts.

For years I fought a losing battle to keep away from involvement with the notion of earthquake prediction. The press and the public will go toward the suggestion of prediction like hog to the trough. Meanwhile, other objects of investigation are neglected or distorted; and aid is given to the people who would like to forget the fact that for public safety we don't need prediction—that earthquake risk could be removed, almost completely, by proper building construction and regulation.

BIBLIOGRAPHY

Preface

Richter, Charles F. Papers. Box 7. California Institute of Technology Archives.

Chapter 1. The Magnitude of the Problem

Fuller, Myron L. *The New Madrid Earthquake.* Washington, D.C.: Government Printing Office, 1912.

Gilbert, Grove Karl. "The Investigation of the California Earthquake of 1906." In *The California Earthquake of 1906.* San Francisco: A. M. Robertson, 1907.

Hernon, Peter. *8.4.* New York: G. P. Putnam's Sons, 1999.

Hough, Susan Elizabeth, and Roger Bilham. *After the Earth Quakes: Elastic Rebound on an Urban Planet.* New York: Oxford University Press, 2005.

Richter, Charles F. Papers. Box 1. California Institute of Technology Archives.

Stein, Seth, and Emile A. Okal. "Speed and Size of the Sumatra Earthquake." *Nature* 434 (2005): 581–82.

Chapter 2. Formative Years

Blount, Jim, Dr. Charles Francis. "Richter Created Scale to Measure Earthquake Magnitude." *Journal News*, February 2, 2000.

"Charles Frances Richter." http://en.wikipedia.org/wiki/Charles_Richter.

"Charles Frances Richter." http://ohiobio.org/richter.htm.

Richter, Charles. *Elementary Seismology.* San Francisco: W. H. Freeman, 1958.

———. Papers. Boxes 1–5. California Institute of Technology Archives.

"Richter, Charles." www.cartage.org.lb/en/themes/Biographies/MainBiographies/R/RichterC/1.html.

Witze, Alexandra. "Charles F. Richter, 1900–1985, American Seismologist." In *Notable Scientists from 1900 to the Present*, ed. Brigham Narins. Farmington Hills, Mich.: Gale Group, 2001.

Chapter 3. Margaret Rose

Cass, Hyla, M.D. "Nutritional Approaches to Mental Health." http://www.ghchealth.com/nutrional-approaches-to-mental-health.html.

Christie, Jonathan. "Diabetes without Complications!" 1998. http://www.survivediabetes.com/hypt2.htm.

Light, Marilyn. *Hypoglycemia: One of the Most Widespread and Misdiagnosed Diseases.* New York: McGraw-Hill, 1999.

Nasar, Sylvia. *A Beautiful Mind: A Biography of John Forbes Nash, Jr.* London: Faber and Faber, 1994.

Richter, Charles F. Papers. Box 1. California Institute of Technology Archives.

Suarez, E. "Relations of Trait Depression and Anxiety to Low Lipid and Lipoprotein Concentrations in Healthy Young Adult Women." *Psychosomatic Medicine* 61 (1999): 273–79.

Woolf, Virginia. *A Room of One's Own.* San Diego: Harvest Books, 1989.

Youngson, R. M., and I. Schott. *Medical Blunders.* New York: New York University Press, 1996.

Zimmer, Gene. "Psychiatric Drugs: Thorazine." 1999. http://www.sntp.net/drugs/thorazine.htm.

Chapter 4. Harnessing the Horses

Einstein, A., B. Podolsky, and N. Rosen. "Can Quantum-Mechanical Description of Physical Reality be Considered Complete?" *Physical Review* 47 (1935): 777–80.

Goodstein, Judith R. *Millikan's School: A History of the California Institute of Technology.* New York: Norton, 1991.

Richter, Charles F. Interview by Ann Scheid, 1979. 156 pp. California Institute of Technology Archives.

———. Papers. Boxes 1, 3, 4. California Institute of Technology Archives.

———. "Quantum Theory of the Spinning Electron." Ph.D. thesis, California Institute of Technology, 1928.

Chapter 5. Earthquake Exploration

Bowen, Ira S. Papers, 45.772. Huntington Library, Pasadena, California.

Dutton, Clarence Edward. "The Charleston Earthquake of August 31, 1886." In *Ninth Annual Report, 1887–88.* Washington, D.C.: U.S. Geological Survey, 1889.

Fuller, Myron L. *The New Madrid Earthquake.* Washington, D.C.: Government Printing Office, 1912.

Goodstein, Judith R. *Millikan's School: A History of the California Institute of Technology.* New York: Norton, 1991.

Hough, Susan E., John G. Armbruster, Leonard Seeber, and Jerry F. Hough. "On the Modified Mercalli Intensities and Magnitudes of the 1811–1812 New Madrid, Central United States, Earthquakes." *Journal of Geophysical Research* 105 (2000): 23839–64.

Hough, Susan Elizabeth, and Roger Bilham. *After the Earth Quakes: Elastic Rebound on an Urban Planet.* New York: Oxford University Press, 2005.

Johnston, Arch C., and Eugene S. Schweig. "The Enigma of the New Madrid Earthquakes of 1811–1812." *Annual Review of Earth and Planetary Sciences* 24 (1996): 339–84.

Lawson, Andrew C. *The California Earthquake of April 18, 1906. Report of the State Earthquake Investigation Commission.* Vol. 1. Washington, D.C.: Carnegie Institution of Washington, 1908; reprinted 1969.

McClellan, Guy R. *The Golden State: A History of the Region West of the Rocky Mountains.* Philadelphia: W. Flint; Chicago: Union Publishing Company, 1872.

Prescott, William, H. "Circumstances Surrounding the Preparation and Suppression of a Report on the 1868 California Earthquake." *Bulletin of the Seismological Society of America* 72 (1982): 2389–93.

Richter, Charles F. Interview by Ann Scheid, 1979. 156 pp. California Institute of Technology Archives.

Wood, Harry Oscar. Papers. Box 112. California Institute of Technology Archives.

Chapter 6. The Kresge Era

Allen, Clarence. "Charles F. Richter: A Personal Tribute." *Bulletin of the Seismological Society of America* 77, no. 6 (1987): 2234–37.

Goodstein, Judith R. *Millikan's School: A History of the California Institute of Technology.* New York: Norton, 1991.

Gutenberg, Hertha. Interview, 1981. Oral History Project, California Institute of Technology Archives.

Kanamori, Hiroo. Response to citation for Walter H. Bucher Medal, 1996. http://www.agu.org/inside/awards/kanamori.html.

Macelwane, James B., ed. *Jesuit Seismological Association, 1925–1950: 25th Anniversary Commemorative Volume.* St. Louis: John S. Swift, 1950.

———. *When the Earth Quakes.* Milwaukee: Bruce, 1947.

Richter, Charles F. Papers. Boxes 1, 23, 30. California Institute of Technology Archives.

Chapter 7. Beno Gutenberg

Allen, Clarence. Interview by David Valone, 1994. Oral History Project, California Institute of Technology Archives.

Goodstein, Judith R. *Millikan's School: A History of the California Institute of Technology.* New York: Norton, 1991.

Gutenberg, Beno, and Charles F. Richter. "Frequency of Earthquakes in California." *Bulletin of the Seismological Society of America* 34, no. 4 (1944): 185–88.

———. *Seismicity of the Earth and Associated Phenomena.* Princeton, N.J.: Princeton University Press, 1949.

Gutenberg, Hertha. Interview by Mary Terrall, February 6 and 13, 1980. 43 pp. California Institute of Technology Archives.

Richter, Charles F. Interview by Ann Scheid, 1979. 156 pp. California Institute of Technology Archives.

Ishimoto, M., and K. Iida. "Observations of Earthquakes Registered with the Microseismograph Constructed Recently." *Bulletin of the Earthquake Research Institute, University of Tokyo* 17 (1939): 443–79.

Knopoff, Leon. "Beno Gutenberg, June 4, 1889–January 25, 1960." Biographical memoirs, National Academy of Sciences. http://www.nap.edu/readingroom/books/biomems/bgutenberg.html.

———. "Beno Gutenberg, June 4, 1889–January 25, 1960." *Deutsche Geophysikalische Gesellschaft e. V.* 4 (1999): 2–15.

Macelwane, James B. Papers. Scrapbooks and photo albums. St. Louis University Archives.

Millikan, Robert. Papers. Box 335. California Institute of Technology Archives.

Oldham, Richard D. "The Constitution of the Interior of the Earth as Revealed by Earthquakes." *Proceedings of the Royal Society,* August 1906, 470.

Press, Frank. Interview, 1983. Oral History Project, California Institute of Technology Archives.

Richter, Charles F. *Elementary Seismology.* San Francisco: W. H. Freeman, 1958.

———. Interview by Ann Scheid, 1979. 156 pp. California Institute of Technology Archives.

———. "An Instrumental Earthquake Scale." *Bulletin Seismological Society of America* 25 (1935): 1–32.

———. Papers. Boxes 1, 23, 26. California Institute of Technology Archives.

Sharp, Robert A. A tribute to Beno Gutenberg. Robert A. Sharp Papers. California Institute of Technology Archives.

———. Papers. Box 6.34. California Institute of Technology Archives.

"They Study Ma Earth's Many Shivers and Moans." *Los Angeles Times.* Front page, October 3, 1929.

Chapter 8. Earthquake!

Hanks, T. C. "The Lompoc, California, Earthquake (November 4, 1927; M = 7.3) and Its Aftershocks." *Bulletin of the Seismological Society of America* 69 (1979): 451–62.

Richter, Charles F. *Elementary Seismology.* San Francisco: W. H. Freeman, 1958.

———. Papers. Boxes 1, 3. California Institute of Technology Archives.

CHAPTER 9. RICHTER SCALE

Aki, Keiti. "Generation and Propagation of G Waves from the Niigata Earthquake of June 16, 1964: Part 2. Estimation of Earthquake Moment, Released Energy and Stress Drop from the G Wave Spectra." *Bulletin Earthquake Research Institute University of Tokyo* 44 (1966): 73–88.

Allen, Clarence. Interview by David Valone, 1994. Oral History Project, California Institute of Technology Archives.

"Charles F. Richter Dies; Earthquake Scale Pioneer." *Los Angeles Times*, September 30, 1985, 1, 18–19.

Drake, Daniel. *Natural and Statistical View, or Picture of Cincinnati and the Miami County, illustrated by maps.* Cincinnati: Looker and Wallace, 1815.

Fuller, Myron L. *The New Madrid Earthquake.* Washington, D.C.: Government Printing Office, 1912.

Gutenberg, Beno, and Charles F. Richter. "Frequency of Earthquakes in California." *Bulletin of the Seismological Society of America* 34, no. 4 (1944): 185–88.

Gutenberg, Hertha. Interview by Mary Terrall, February 6 and 13, 1980. 43 pp. California Institute of Technology Archives.

Hanks, T. C., and H. Kanamori. "A Moment Magnitude Scale." *Journal of Geophysical Research* 84 (1979): 2348–50.

Hernon, Peter. *8.4.* New York: G. P. Putnam's Sons, 1999.

Hough, Susan E., John G. Armbruster, Leonard Seeber, and Jerry F. Hough. "On the Modified Mercalli Intensities and Magnitudes of the 1811–1812 New Madrid, Central United States, Earthquakes." *Journal of Geophysical Research* 105 (2000): 23839–64.

Ishimoto, M., and K. Iida. "Observations of Earthquakes Registered with the Microseismograph Constructed Recently." *Bulletin of the Earthquake Research Institute, University of Tokyo* 17 (1939): 443–79.

Lay, Thorne, H. Kanamori, C. J. Ammon, M. Nettles, S. N. Ward, R. C. Aster, S. L. Beck, S. L. Bilek, M. R. Brudzinski, R. Butler, H. R. DeShon, G. Ekstrom, K. Satake, and S. Sipkin. "The Great Sumatra-Andaman Earthquake of 26 December 2004." *Science* 308 (2005): 1127–33.

McMurtrie, H., M.D. *Sketches of Louisville and its Environs, Including, among a great variety of miscellaneous matter, a Florula Louisvillensis; or, a catalogue of nearly 400 genera and 600 specied of plants, that grow in the vicinity of the town, exhibiting their generic, specific, and vulgar English names.* Louisville: S. Penn., Junior, 1839.

Press, Frank. Interview, 1983. Oral History Project, California Institute of Technology Archives.

Richter, Charles F. *Elementary Seismology.* San Francisco: W. H. Freeman, 1958.

———. "An Instrumental Earthquake Scale." *Bulletin Seismological Society of America* 25 (1935): 1–32.

———. Interview by Ann Scheid, 1979. 156 pp. California Institute of Technology Archives.

———. Papers. Boxes 1, 23, 24, 25, 34. California Institute of Technology Archives.

Thatcher, Wayne. "Strain Accumulation and Release Mechanism of the 1906 San Francisco Earthquake." *Journal Geophysical Research* 80 (1975): 4862–72.

Wadati, K. "Shallow and Deep Earthquakes. 3rd paper. *Geophysical Magazine* 4 (1931): 231–85.

Wald, David J., Hiroo Kanamori, and Donald V. Helmberger. "Source Study of the 1906 San Francisco Earthquake." *Bulletin of the Seismological Society of America* 83 (1993): 891–1019.

Chapter 10. Charlie

Allen, Clarence. "Charles F. Richter: A Personal Tribute." *Bulletin of the Seismological Society of America* 77, no. 6 (1987): 2234–37.

"Charles F. Richter Dies; Earthquake Scale Pioneer." *Los Angeles Times,* September 30, 1985, 1, 18–19.

Gutenberg, Beno. Papers. Box 3.10. California Institute of Technology Archives.

Holliday, Kate. "Backpacking: The Ageless Art." *Field and Stream,* 58–59, 136.

Kaufman, Robert. "It's Really the Gutenberg-Richter Scale." Letter, *New York Times,* November 23, 1989.

Memorial service for Charles Richter, November 14, 1985. Audiotape. California Institute of Technology Archives.

Richter, Charles F. *Elementary Seismology.* San Francisco: W. H. Freeman, 1958.

———. Interview by Ann Scheid, 1979. 156 pp. California Institute of Technology Archives.

———. "Our Earthquake Risk—Facts and Non-Facts." *Engineering and Science,* January 1964.

———. Papers. Boxes 1, 2–3, 5–7, 8, 15, 23–24, 26, 27, 29. California Institute of Technology Archives.

Richter, Charles F., and Beno Gutenberg. Interview by CBS News, 1950s. Tape recording. California Institute of Technology Archives.

Symposium, 50th Anniversary of Seismological Laboratory. Tape recording. California Institute of Technology Archives.

Chapter 11. Lillian

Brand, Lillian, "Writing is Fun." *Recreation*, February 1949, 499.
Broderson, Gladys Virginia. Scrapbook of Grand Canyon, Arizona, 1937.
"Charles F. Richter Dies; Earthquake Scale Pioneer." *Los Angeles Times*, September 30, 1985, 18–19.
Chrohn's and Colitis Foundation, Peter A. Banks, Daniel H. Prescott, and Penny Steiner, eds. *The Crohn's Disease and Ulcerative Colitis Fact Book*. New York: Hungry Minds, 1983.
Cinder, Cec. *The Nudist Idea*. Riverside, Calif.: Ultraviolet Press, 1998.
Elysia, Valley of the Nude. Directed by Brian Foy, 1933.
"Famed Nudist Richter's Personal Legacy Lost in Quake." *Nude and Natural*, spring 1994, 6.
"Fraternity Elysia." *The Nudist*, January 1935, 29.
Glassey, Hobart. "Building at Elysian Fields." *The Nudist*, November 1934, 19–20.
———. "Elysia at Elsinore." *The Nudist*, March 1934, 15, 18–19.
———. "Getting on Good Terms with the Press." *The Nudist*, January 1935, 20–21.
Gutenberg, Hertha. Interview by Mary Terrall, February 6 and 13, 1980. 43 pp. California Institute of Technology Archives.
"House for Dr. Charles Richter, Pasadena, California, Richard J. Neutra, Architect, Peter Pfisterer, Collaborator." *Architectural Forum*, 1937, 214–15.
Pasadena Star News. Obituary, Lillian Richter, November 6, 1972.
Richter, Charles F. Interview by Ann Scheid, 1979. 156 pp. California Institute of Technology Archives.
———. Papers. Boxes 1, 2, 7, 8, 29. California Institute of Technology Archives.
Saunders, Lilian. "Mist." *Poet Lore*, December 1925, 629.
———. "Moonlight." *Poet Lore*, December 1925, 630.
———. "Sea Jewels." *Poet Lore*, December 1925, 630.
———. "Seven O'Clock." *Poet Lore*, September 1926, 431.
Simmons, Elaine. "Richters Remembered. *Pasadena Star News*, October 7, 1985.
Storey, Mark. *Cinema au Naturel: A History of Nudist Film*. Oshkosh, Wis.: Naturist Educational Foundation, 2003.
Sunshine and Health, June 1948, 6.
Unashamed. Directed by Allen Stuart, 1938.
Woolf, Virginia. *A Room of One's Own*. San Diego: Harvest Books, 1989.

Chapter 12. Richter's Women

Hudson, Glenda A. *Sibling Love and Incest in Jane Austen's Fiction*. London: Palgrave Macmillan, 1999.

Richter, Charles F. Papers. Boxes 1, 2, 7, 25. California Institute of Technology Archives.

Winchester, Simon, *A Crack in the Edge of the World: America and the Great California Earthquake of 1906*. New York: HarperCollins, 2005.

Chapter 13. Autumn

Fowke, Edith, and Joe Glazer, eds. *Songs of Work and Protest*. New York: Dover, 1973.

Harrison, Henry, ed. *California Poets: An Anthology of 244 Contemporaries*. New York: Henry Harrison, 1932.

McClintok, Harry. "Hallelujah! I'm a Bum." http://unionsong.com/u029.html.

Richter, Charles F. *Elementary Seismology*. San Francisco: W. H. Freeman, 1958.

———. Papers. Boxes 1–4, 7, 8. California Institute of Technology Archives.

Storey, Mark. *Cinema au Naturel: A History of Nudist Film*. Oshkosh, Wis.: Naturist Educational Foundation, 2003.

Chapter 14. Asperger's Syndrome

Asperger, Hans. "Die 'autistischen psychopathen' im kindesalter." *Archiv fur Pyschiatrie und Nervenkankheiten* 117 (1944): 76–136.

Attwood, Tony. *Asperger's Syndrome: A Guide for Parents and Professionals*. London: Jessica Kingsley, 1998.

———. "The Pattern Abilities and Development of Girls with Asperger's Syndrome." *The Source*, September 1999.

Atwood, Margaret. *Cat's Eye*. New York: Doubleday, 1989.

Bashe, Patricia Romanowski, and Barbara L. Kirby. *The OASIS Guide to Asperger Syndrome: Advice, Support, Insight, and Inspiration*. New York: Crown, 2001.

Broderson, Gladys Virginia. Scrapbook of Grand Canyon, Arizona, 1937.

"Charles F. Richter Dies; Earthquake Scale Pioneer." *Los Angeles Times*, September 30, 1985, 18–19.

Frith, U., ed. *Autism and Asperger Syndrome*. Cambridge: Cambridge University Press, 1998.

Gutenberg, Hertha. Interview by Mary Terrall, February 6 and 13, 1980. 43 pp. California Institute of Technology Archives.

Holliday, Kate. "Backpacking: The Ageless Art." *Field and Stream*, February 1973, 58–59, 136.

———. "Charles Richter, Earthquake Man." *Smithsonian*, February 1971, 69–72.

Jackson, Luke, *Freaks, Geeks, and Asperger Syndrome: A User Guide to Adolescence*. London: Jessica Kingsley, 2002.

Kalb, Claudia. "When Does Autism Start?" *Newsweek*, February 28, 2005, 45–53.
Memorial service for Charles Richter, November 14, 1985. Audiotape. California Institute of Technology Archives.
Palmer, Raymond F., Steven Blanchard, Zachary Stein, David Mandell, and Claudia Miller. "Environmental Mercury Release, Special Education Rates, and Autism Disorder: An Ecological Study of Texas." *Health and Place* 12 (June 2006): 203–9.
Richter, Charles F. *Elementary Seismology.* San Francisco: W. H. Freeman, 1958.
———. Interview by Ann Scheid, 1979. 156 pp. California Institute of Technology Archives.
———. Papers. Boxes 1–5, 7–9, 31, 34. California Institute of Technology Archives.
———. "A Seismologist in Japan." *Engineering and Science*, March 1961, 24–30.
Richter, Charles F., and Beno Gutenberg. Interview by CBS News, 1950s. Tape recording. California Institute of Technology Archives.
Sainsbury, Clare. *Martian in the Playground: Understanding the Schoolchild with Asperger's Syndrome.* Lucky Duck Publications, 2000.
Taylor, B., et al. "MMR Vaccine and Autism: No Epidemiological Evidence for a Causal Association." *The Lancet* 363 (2004): 747.
Wakefield, A. J., et al. "Ileal Lymphoid Nodular Hyperplasia, Non-specific Colitis, and Regressive Developmental Disorder in Children." *The Lancet* 351 (1998): 637–41.
Willey, Liane, *Pretending to be Normal: Living with Asperger's Syndrome.* London: Jessica Kingsley, 1999.
Williams, Karen. "Understanding the Student with Asperger Syndrome: Guidelines for Teachers." *Focus on Autistic Behavior* 10, no. 2 (June 1995).

Chapter 15. Here It Comes Again

Gomberg, J., and P. Bodin. "Triggering of the Ms = 5.4 Little Skull Mountain Nevada, Earthquake with Dynamic Strains." *Bulletin of the Seismological Society of America* 84 (1994): 844–53.
Hill, David, et al. "Seismicity in the Western United States Remotely Triggered by the M7.4 Landers, California, Earthquake of June 28, 1992." *Science* 260 (1993): 1617–23.
King, Geoffrey C. P., and David D. Bowman. "The Evolution of Regional Seismicity between Large Earthquakes." *Journal of Geophysical Research* 108 (2003), doi:10.1029/201JB000783.
Richter, Charles F. "Here It Comes Again." *Natural History*, 1972.
———. Interview by Ann Scheid, 1979. 156 pp. California Institute of Technology Archives.

Richter, Charles F. Papers. Boxes 3, 7, 28. California Institute of Technology Archives.

Stein, Ross. "The Role of Stress Transfer in Earthquake Occurrence." *Nature* 402 (1999): 605–9.

Chapter 16. Predicting the Unpredictable

Alexander, George. "Can We Predict the Coming California Quake?" *Popular Science*, November 1976, 79–82.

Allen, C. R., P. St. Armand, C. F. Richter, and J. M. Nordquist. "Relationship between Seismicity and Geologic Structure in the Southern California Region." *Seismological Society of America* 55 (1965): 753–97.

Allen, Clarence R. Interview by Stanley Scott, 2002. 119 pp. Connections, EERI Oral History Series, Oakland, California.

Bawden, G. W., A. Donnellan, L. H. Kellogg, D. N. Dong, and J. B. Rundle. "Geodetic Measurements of Horizontal Strain Near the White Wolf Fault, Kern County, California, 1926–1993." *Journal of Geophysical Research* 102 (1997): 4957–67.

Castle, R. O., J. P. Church, and M. R. Elliott. "Aseismic Uplift in Southern California." *Science* 192 (1976): 251–53.

Gutenberg, Beno, and Charles F. Richter. "Frequency of Earthquakes in California." *Bulletin of the Seismological Society of America* 34, no. 4 (1944): 185–88.

Hill, Robert T. *Southern California Geology and Los Angeles Earthquakes.* Los Angeles: Southern California Academy of Sciences, 1928.

"In His Own Words: Just Like the Movie, a Young Scientist Predicts an L.A. Earthquake." *People*, May 17, 1976, 49–55.

International News Service Press Release, Stanford University, March 1933.

Ishimoto, M., and K. Iida. "Observations of Earthquakes Registered with the Microseismograph Constructed Recently." *Bulletin of the Earthquake Research Institute, University of Tokyo* 17 (1939): 443–79.

Jackson, David D., Wook B. Lee, and Chi-Ching Liu. "Height Dependent Errors in Southern California Leveling." In *Earthquake Prediction: An International Review*, ed. David W. Simpson and Paul G. Richards. Washington, D.C.: American Geophysical Union, 1981.

Keilis-Borok, V., P. Shebalin, K. Aki, A. Jin, A. Gabrielov, D. Turcotte, Z. Liu, and I. Zaliapin. "Documented Prediction of the San Simeon Earthquake 6 Months in Advance: Premonitory Change of Seismicity, Tectonic Setting, Physical Mechanism." Abstract. Annual Meeting, Seismological Society of America, Palm Springs, California, 2004.

Kerr, R. A. "Palmdale Bulge Doubts Now Taken Seriously." *Science* 214 (1981): 1331–33.

Nadeau, Robert M., and David Dolenc. "Nonvolcanic Tremors beneath the San Andreas Fault." *Science* 307 (2005): 389.

Obara, K. "Nonvolcanic Deep Tremor Associated with Subduction in Southwest Japan." *Science* 296 (2002): 1679–81.

Reilinger, Robert, and Larry Brown. "Neotectonic Deformation, Near-Surface Movements and Systematic Errors in U.S. Releveling Measurements: Implications for Earthquake Prediction." In *Earthquake Prediction: An International Review,* ed. David W. Simpson and Paul G. Richards. Washington, D.C.: American Geophysical Union, 1981.

Richter, Charles F. Interview by Harry Spall, U.S. Geological Survey, Reston, Virginia, 1980. http://neic.usgs.gov/neis/seismology/people/int_richter.html.

———. Papers. Boxes 23, 26, 28, 30. California Institute of Technology Archives.

Stein, Ross. "Discrimination of Tectonic Displacement from Slope-Dependent Errors in Geodetic Leveling from Southern California. In *Earthquake Prediction: An International Review,* ed. David W. Simpson and Paul G. Richards. Washington, D.C.: American Geophysical Union, 1981.

Stein, R. S., C. T. Whalen, S. R. Holdahl, and W. E. Strange. "Saugus-Palmdale, California, Field-Test for Refraction Error in Historical Leveling Surveys." *Journal of Geophysical Research* 91 (1986): 9031–44.

Chapter 17. Sizing Up Earthquake Hazard

Atwater, Brian F., and D. K. Yamaguchi. "Sudden, Probably Coseismic Submergence of Holocene Trees and Grass in Coastal Washington State." *Geology* 19 (1991): 706–9.

Diehl, Digby. *Super-Talk.* Garden City, N.Y.: Doubleday, 1974.

Heaton, Thomas H., and Hiroo Kanamori. "Seismic Potential Associated with Subduction in the Northwestern United States." *Bulletin of the Seismological Society of America* 74, no. 3 (1984): 933–41.

Richter, Charles F. *Elementary Seismology.* San Francisco: W. H. Freeman, 1958.

———. "Here It Comes Again." *Natural History,* 1972.

———. Papers. Boxes 3, 26. California Institute of Technology Archives.

Sharp, Robert P. Interview, 1981. Oral History Project, California Institute of Technology Archives.

Chapter 18. Hazard in a Nuclear Age

Committee on Opportunities for Accelerating Characterization and Treatment of Waste at DOE Nuclear Weapon Sites. *Improving the Characterization and Treatment of Radioactive Wastes for the Department of Energy's Accelerated Site Cleanup Program.* Washington, D.C.: National Academies Press, 2005.

Nuclear Regulatory Commission. Hearings transcripts, July 20–21, 1976. Richter, Charles F. Papers. Box 9. California Institute of Technology Archives.

CHAPTER 19. SUPERNOVA

Allen, Clarence. "Charles F. Richter: A Personal Tribute." *Bulletin of the Seismological Society of America* 77, no. 6 (1987): 2234–37.

———. Interview by David Valone, 1994. Oral History Project, California Institute of Technology Archives.

———. Interview by Stanley Scott, 2002. 119 pp. Connections, EERI Oral History Series, Oakland, California.

Barbour, John. "Charles Richter." *Los Angeles Times Home*, May 11, 1980.

———. "Creator of Richter Scale Knows Fallibility of Earth." *San Gabriel Valley Tribune*, January 25, 1981, A10–A11.

Barlow, H. L. 1933 Long Beach Earthquake Scrapbook. Ephemera collection, no. 38, Huntington Library, Pasadena, California.

"Earthquake Scale Inventor Near Death." *Los Angeles Times*, August 21, 1984.

Forrest, Michael R. "Charles Richter, Part Two." *Southern California Earthquake Center*, v. 4, (1998): 26–29.

Gutenberg, Beno. Papers. Box 3.10. California Institute of Technology Archives.

Hewitt, Jerene. *Pipes of Pan* (Pasadena City College), 1972, 22, 42, 72.

Isacks, Brian, Jack Oliver, and Lynn R. Sykes. "Seismology and the New Global Tectonics." *Journal of Geophysical Research* 73 (1968): 5855–99.

Memorial service for Charles Richter, November 14, 1985. Audiotape. California Institute of Technology Archives.

Richter, Charles F. Comments on paper by B. Isacks, J. Oliver, and L. R. Sykes, "Seismology and New Global Tectonics." *Journal of Geophysical Research* 74 (1969): 2786–88.

———. "Earthquake Light in Focus." *Science* 194 (1976): 259.

———. Papers. Boxes 1, 2, 7, 9, 24, 26, 30, 33, 34. California Institute of Technology Archives.

———. "Possible Seismicity of Nacimiento Fault, California." *Geological Society of America Bulletin* 80 (1969): 1363–70.

———. Retirement letter. Caltech Seismology Laboratory, archives.

———. "Transverse Aligned Seismicity and Concealed Structures." *Science* 166 (1969): 173–78.

Tucker, Wanda. "Richter: A Down-to-Earth Friend." *Pasadena Star News*, October 3, 1985.

Wilford, Jean Notle, "Charles Richter, Quake Expert, Dies." *New York Times*, October 1, 1985.

APPENDIX. A BELATED FAREWELL

Richter, Charles F. Papers. Box 9. California Institute of Technology Archives.

ACKNOWLEDGMENTS

Sometimes one doesn't know where to begin with acknowledgments, but in this case it is easy: this story could not have been written had Charles Richter not made the decision to leave his papers to the Caltech archives. Those archives are today tended by archivist Dr. Judith Goodstein and her most capable staff, to whom I am eternally indebted; especially Bonnie Ludt, Shelley Erwin, and Kevin Knox. When I first stumbled down into the deepest reaches of the Beckman Hall subbasement, around corners, and past doors bearing signs warning, "Danger—Laser Radiation," I had no idea what to expect. What I found was not only a treasure trove of information but also the wonderful group of individuals in whose hands this information, and so much more, has been entrusted for safekeeping. I also thank John Waide at the St. Louis University archives for a lovely and enormously helpful afternoon, and the staff at the Huntington Library in San Marino, in particular Romaine Ahlstrom and Daniel Lewis, for their help. It is hard to imagine a more hospitable environment to do research than the Huntington, but I admit that I have never been more proud to be a U.S. taxpayer than I was during the hours that I spent at the Library of Congress—truly a national treasure.

Richter's decision to leave his papers, including considerable personal material, to the archives means that he is the definitive authority for much of the information in this book. The story could not have been told, however, had it not been for the recollections of those who were personally acquainted with Charles or Lillian Richter—recollections that many people have graciously shared with me. My conversations naturally began close to home, with colleagues of Richter's who have remained at Caltech: fellow seismologists Clarence Allen, Hiroo Kanamori, Don Anderson, Tom Heaton, and Kate Hutton, and senior instrumentation specialists Dave Johnson and Bob Taylor. I am especially indebted to Johnson as well as Clarence Allen, one of Richter's closest longtime colleagues, for their help and patience, and for many thoughtful, constructive comments on the manuscript. (Allen now has the unique distinction of having contributed to manuscript preparations for both Richter's own book and his biography.)

From the corridors of Caltech the trail led naturally to colleagues who had been at the Seismo Lab during Richter's tenure and since moved on to establish careers at other institutions. This list includes a number of former students: Shelton Alexander, Tom Jordan, Bob Geller, Tom Hanks, John Gardner, David Hill, Barry Keller, John Lett, Alexander Goetz, Harmut Spetzler, Ta-Liang Teng, Emile Okal, Bob Phinney, and Nafi Toksoz, as well as Betty Shor, who in the 1950s was not only the wife of Richter's graduate student George Shor but also Richter's paid assistant. To my surprise, I found that no matter how much I had learned previously about Richter and the Seismo Lab, every conversation added something new to the portrait—a bit of color, an interesting new detail. I learned how important seemingly small bits of information—Tom Jordan's "bang the Wang" story comes to mind—could be in bringing a story, and a man, to life.

Having sadly missed my chance to talk with Vi Taylor, I am especially indebted to Betty Shor for sharing her unique perspective and experiences. It was only in talking with her that I realized that some of the conventional wisdom that persists to this day, about Lillian Richter in particular, has been colored significantly by conventions—social conventions of a bygone era when faculty wives were expected to fit a certain mold. Lillian Brand Richter broke that mold with a vengeance. When I first began researching the story, I knew Lillian only from the limited glimpses that Richter's colleagues had of her during her lifetime. By the time I finished she had become for me what she became for Richter later in his life: so much more of a real person, one whose life story was compelling in its her own right.

Other of Richter's former colleagues shared their recollections with me as well: Freeman Gilbert, Jim Brune, Frank Press, Leon Knopoff, Bob Taylor, Lynn Sykes, Eric Lindvall, and Karen McNally. One of the many pleasures of writing this book has been the chance to get to know a few colleagues with whom I had not been previously acquainted—although many I had been happy to count among my own colleagues and friends. It is sobering to read over the list of scientists with whom I have spoken: it is nothing less than a Who's who of American seismology. Eric Lindvall's recollections were especially helpful to flesh out the portrait of Richter's later years, after his retirement from Caltech. Conversations with Karen McNally were especially illuminating as well: having arrived at the Seismo Lab as a postdoc in 1976, she was the last seismologist to work closely with Richter in an academic setting.

I am further indebted to the current director of the Seismo Lab, Jeröen Tromp, for assistance with the project; also to Viola Carter, Elisa Weffor, and Rosemary Miller.

Richter's nuclear family might have been especially nuclear, but I was fortunate to find Lillian's niece and nephew, Dorothy Crouse and Bruce Walport, now of Oxnard and Grants Pass, Oregon, respectively. Other family members graciously shared their recollections as well: Laurie Walport, Mary White, and Kathy Haag. Of all the people I spoke with, they alone had known "Uncle Charles" and "Aunt Lil" for their entire lives; they alone could fill in some of my most vexing blanks about the Richters' personal lives.

I have met a couple of other people along the way whose recollections were tremendously helpful, including the Richters' neighbor on Villa Zanita, Lawrence Fusha. Without question, my two most serendipitous sources would have to be Warren Boehm and Bryan Mead. Boehm materialized one day out of a clear blue sky, a member of a local historical society who had happened to sign up for a fault tour that I was leading. At a time when I was enormously frustrated by my inability to find out about Lillian's writing career, up popped Boehm. After finishing his college degree in the early 1960s, had decided to take a writing class for the fun of it—and ended up in an evening class at Glendale Junior College taught by Lillian Brand. Mead materialized by way of his response to a query I had posted on an online bulletin board—he thought I might be interested in the marvelous scrapbook he had just purchased at a local flea market. One boggles at the idea that casual conversations during a chance encounter between strangers seventy-five years ago could, first, end up preserved for posterity, and, second, find their way to someone who is looking for them. Of course such things don't "just happen": they happen because there are people, like Mead, who keep history alive.

I must also express my gratitude to a number of individuals, some of whom did not know Charles Richter, but all of whom became part of my journey in putting this story together. At the head of this list is my father, Jerry Hough, who probably never imagined that his daughter actually absorbed the real lesson of visiting the Duke University library to read original political cartoons for a tenth-grade history report. Somewhat later in life he showed me the ropes at the Library of Congress, and I believe is amused to know that I put these lessons to work tracking down nudist magazines from the 1930s (and a few more staid publications).

Two of my colleagues went above and beyond the call of duty, reviewing the manuscript, passing along comments, and calling my attention to materials that I had overlooked. To Cliff Thurber and Seth Stein, thank you, thank you, thank you. I am further indebted to Ross Stein, David Jackson, Nancy King, and Ken Hudnut for relaying their experiences during the Palmdale Bulge affair; also to Roger Bilham, for research assistance when I discovered that the library at his university—but not Caltech—has old issues of *Poet Lore.*

My research led me in a number of interesting directions, which in turn led me to a great many interesting and enormously helpful people: Edna Shalev at *Field and Stream*, Caroyln Payne at Glendale Community college, the reference librarians at Pasadena City College, staff members at the Nuclear Regulatory Commission, Amy Ulmer at PCC, the staff of Arundel Books in Los Angeles, Jason DeYoung with The Writer's Center, Carolyn Nash at the Mountain View Cemetery, Flo Nilson of Olympic Fields, Brad McNeil at the Pasadena Historical Museum, former teacher and realtor (and current writer and artist) Elaine Simmons, and David Perlman at the *San Francisco Chronicle.* (That this is an interesting list says something about the man whose life I was researching.) Others who helped along the way with advice and suggestions include Alan Lamson, Susan Lehman, Sal Towse, and Kevin Kilty; also Mark Storey and Cec Cinder, who graciously shared their knowledge of the early nudist movement in Southern California.

The actual research and writing are, of course, small parts of the process by which books are born. I thank Simon Winchester for his advice and support, and for possibly knowing before I did that this was the book I was born to write (although, dear Simon, we do need to talk about that footnote). I am also indebted to Simon for introducing me to Alexis Hurley, my agent at Inkwell, now a confirmed Friend of Charlie. Little is as gratifying to a writer as the opportunity to work with people who share one's sense of excitement and optimism for a book. In this case I have been doubly blessed to have worked with both Alexis and my brilliant and wonderful editor at Princeton University Press, Ingrid Gnerlich. Ingrid also knew things about the book before I did, namely that Princeton truly was the right place for Charlie and me.

On a more personal level I am indebted to a number of people who are part of my life, starting with a group of individuals who have taught me what it means to "quiet the mind," and "explore one's boundaries": Patsy, Jodi, Constance, and other friends at Mission Street Yoga; Nancy and Kelley at the Caltech gym; Naader, David, Meg, and John at Yoga Kingdom sanctuary. To work at a research position all day and come home and write in the evening, especially a story as complex as this one, requires an energy and focus that I'm not sure I would have had if my friend and neighbor

Barbara Kabaelo hadn't first invited me to join her in a class one Saturday. (Thank you, Barbara!)

I can also thank my husband, Lee Slice, for support. Researching this book did not require much travel, but writing it required a great many antisocial hours in my study. I do believe his appreciation for seismology research as an interesting field of inquiry grew by leaps and bounds with this project. Specifically, that would have been the day that my VHS copies of *Elysia* and *Unashamed* arrived in the mail. I would also thank my children, Sarah, Joshua, and Paul, for support and encouragement, but truth in advertising prevents this. It does, however, continue to be a pleasure to watch them grow into capable, resourceful, and usually rational older teens and young adults.

The person to whom I am the most deeply indebted is the one whom I never had a chance to meet, and whom I cannot properly thank. I believe he wanted his story told some day; that he wanted the world to know the man behind the scale. I went into this story knowing only that I wanted to learn more about Richter's remarkable, and a remarkably complex, life. Every answer that I found led to more questions; every lead that I chased seemed to catapult me off in new directions. It is possible, I know, that Richter had secrets he did not choose to write about; that he did not wish to share. I can respect this. In the end I didn't find the answer to every question. I hope I found enough.

INDEX

EARTHQUAKES BY DATE